D0893393

The Virtues of Ignorance

Culture of the Land

A Series in the New Agrarianism

This series is devoted to the exploration and articulation of a new agrarianism that considers the health of habitats and human communities together. It demonstrates how agrarian insights and responsibilities can be worked out in diverse fields of learning and living: history, science, art, politics, economics, literature, philosophy, religion, urban planning, education, and public policy. Agrarianism is a comprehensive worldview that appreciates the intimate and practical connections that exist between humans and the earth. It stands as our most promising alternative to the unsustainable and destructive ways of current global, industrial, and consumer culture.

The Virtues of Ignorance

Complexity, Sustainability, and the Limits of Knowledge

EDITED BY BILL VITEK AND WES JACKSON

THE UNIVERSITY PRESS OF KENTUCKY

Copyright © 2008 by The University Press of Kentucky except as noted below.

"Introduction: Taking Ignorance Seriously" copyright © 2008 by Wes Jackson.

"The Way of Ignorance" copyright © 2005 by Wendell Berry from *The Way of Ignorance*. Reprinted by permission of Shoemaker & Hoard Publishers. All other rights remain with the author.

"Toward an Ignorance-Based World View," copyright © 2005 by Wes Jackson, first appeared, in slightly different form, in *The Land Report* 81 (Spring 2005) and is reprinted here courtesy of the author.

"Toward an Ecological Conversation" first appeared in the online newsletter *NetFuture* 127 (January 10, 2002) and was reprinted in the *New Atlantis* as "A Conversation with Nature" (Fall 2003). It is reprinted here, in slightly different form, courtesy of the author.

The University Press of Kentucky

Scholarly publisher for the Commonwealth,
serving Bellarmine University, Berea College, Centre College of Kentucky, Eastern Kentucky University, The Filson Historical Society, Georgetown College, Kentucky Historical Society, Kentucky State University, Morehead State University, Murray State University, Northern Kentucky University, Transylvania University, University of Kentucky, University of Louisville, and Western Kentucky University.
All rights reserved.

Editorial and Sales Offices: The University Press of Kentucky
663 South Limestone Street, Lexington, Kentucky 40508-4008
www.kentuckypress.com

All books in the Culture of the Land series are printed on acid-free recycled paper meeting the requirements of the American National Standard for Permanence in Paper for Printed Library Materials.

12 11 10 09 08   5 4 3 2 1

Library of Congress Cataloging-in-Publication Data

The virtues of ignorance : complexity, sustainability, and the limits of knowledge / edited by Bill Vitek and Wes Jackson.
     p. cm.— (Culture of the land)
  Includes bibliographical references and index.
  ISBN 978-0-8131-2477-3 (hardcover : alk. paper)
  1. Agriculture—Philosophy. 2. Sustainable agriculture. 3. Sustainable development. 4. Philosophy and science. 5. Nature and civilization. 6. Conservation of natural resources. 7. Ignorance (Theory of knowledge) I. Vitek, William, 1957- II. Jackson, Wes. III. Series.
  S494.5.P485V57 2008
  601—dc22   2007048494

Manufactured in the United States of America.

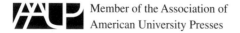

Member of the Association of
American University Presses

Contents

ACKNOWLEDGMENTS

The editors wish to acknowledge and thank the many folks who worked behind the scenes and welcomed us to Matfield Green, Kansas, in the summer of 2004 for the meeting that resulted in this book as well as those whose work is not present here but whose participation enriched the gathering: Ronnie and Tom Armstrong, Ben Champion, Toots Conley, Robert Day, Dennis Dimick, Liz Granberg, Gayle Lasater, Dan Lesh, David Orr, Donna Prizgintas, Sister Helen Ridder, Virginia Smith, Clara Jo, Phyllis and Crystal Talkington, Arion Thiboumery, Kim and Travis Todd, and Fred Whitehead.

The editors also thank Steve Wrinn and his expert staff and editorial reviewers at the University Press of Kentucky for their care and attention at every step of the publication process.

Bill Vitek acknowledges Clarkson University, the Center for Humans and Nature, and the Land Institute for their support of his work in 2007 to bring this book to press.

Wes Jackson acknowledges Darlene Wolf, Jerry Glover, and his staff at the Land Institute.

INTRODUCTION

Taking Ignorance Seriously

Bill Vitek and Wes Jackson

> How can we remember our ignorance, which our growth requires,
> when we are using our knowledge all the time?
>   —Henry David Thoreau

THE QUESTION

Since we're billions of times more ignorant than knowledgeable, why
not go with our long suit and have an ignorance-based worldview?

A few years ago, some well-known scientists published a paper,
followed by a book, in which they assigned a dollar value to nature's
services.[1] The exercise doubtlessly has increased awareness of what ordinary accounting does not count, but we have no idea how such calculations can be reasonably made. We don't even know the full role of any
of the species that have been discovered, let alone those not discovered
or never to be discovered. And then there are the physical forces at work
on our behalf—global climate, for example—some of which we have
clearly altered by our presence. This effort to assign a dollar value represents the current zenith of Enlightenment thought.

It is our contention that a knowledge-based worldview lies at the
very center of the Enlightenment perspective—our perspective—and
both makes possible and drives the pursuits and principles that typically
get all the attention: individual freedom, economic growth, scientific
progress, and the rejection of thermodynamic, material, and moral limits. In a word, the Enlightenment perspective is all about liberty. But

the central footing on which liberty rests and draws nourishment is a knowledge-based worldview.

This worldview has a long history and many sources, from the theft of knowledge in the Genesis creation story or the theft of fire by Prometheus to the Greek emphasis on the powers of human rationality. But it is the Enlightenment that most successfully combined the scholarly pursuits of mathematics, science, and philosophy with the tools of engineers, architects, and physicians. It is what historians describe as the merger of *techne* (the everyday knowledge gained by experience and repetition with little regard for how and why) and *episteme* (the knowledge that comes from the rational pursuit of causes and first principles).

This perspective is mirrored and articulated in the work of many progenitors, beginning in sixteenth-century Europe, who, if we are feeling generous, may be excused for mistaking nature as infinite and infinitely malleable when humans were a scarce, weak species pursing their projects in the small clearings that culture made on our very sizable planet. From their vantage, Johannes Kepler, Nicolaus Copernicus, Michael Servetus, Benedetto Castelli, Phillip von Hohenheim (Paracelsus), Galileo Galilei, John Locke, Thomas Hobbes, Francis Bacon, Baruch Spinoza, Isaac Newton, Voltaire, Pierre Bayle, Charles de Montesquieu, and others could scarcely anticipate the problems of scale that would arise when their ideas and programs were amplified into a human culture weighing in at 7 billion souls. Many of them did their work in Italy, England, Scotland, the Netherlands, and France, where social and intellectual conditions allowed thinkers to more easily begin to break free from the grips of Scholasticism, Aristotelianism (particularly in science), and the authoritarian powers of church and state. Each of these thinkers provided central pieces of a knowledge-based worldview and laid the groundwork for the revolutions to come.

One of the earliest and most fundamental sources of Enlightenment thinking is the work of French philosopher and mathematician René Descartes. Descartes challenged himself to nothing less than putting the human capacity to know the world on an entirely new and—he hoped—foolproof philosophical footing. His work is emblematic of Immanuel Kant's claim that the motto of the Enlightenment should be "Dare to Know." Descartes describes his discoveries as a daylong meditation in a stove-heated room. His subject matter was dreams, the world, God, and, most important, the ability of the individual human mind first to doubt all of it and then to reconstruct—on its own terms—every bit of it in a way that would guarantee truth. Descartes' reward, and to date the modern world's reward, is a knowledge-based bulwark centered on individualized human consciousness and, with it, the ability to make and unmake the world without limits.

This Cartesian moment helped make possible the three revolutions that have been identified with the Enlightenment—scientific, political, and economic—as well as the many revolutionaries who came after and who made the modern world fully operational. Together these revolutions freed cultures to embark on pursuits that were heretofore forbidden or considered impossible: the control of nature; the creation of economies and technologies that went beyond subsistence; the freedom of individuals from governments, religious and family traditions, and the past; and a belief in human progress that is separate from evolution and unencumbered by moral and spiritual beliefs. Descartes' *Meditations on First Philosophy* (1641) is a keystone in a knowledge-centered belief system that insists that the world be remade in humanity's interests.

In trying to overturn the false optimism and many errors, injustices, and ecological disasters that have resulted from this system, an impressive array of critics—from Marxists, human rights activists, and spiritual leaders to environmentalists and indigenous peoples—has attacked one or more of the usual offenders: the extractive/polluting economy, the injustices of capitalism, the mistreatment of animals, and the specter of running out of oil, for example. But, each time an attack comes, the Western defenders rely on Descartes' ace. Yes, we are told, there may be injustices in the world, or leaky factories, or shrinking oil and freshwater reserves, and abuse of animals. But the thinking mind—especially when linked with other thinking minds (the more the better!)—will overcome all these limits and problems. Just a little more time, or another Einstein or Edison or Salk or Gandhi or Fritz Haber and Carl Bosch, and the world will be made better for the world's humans—and perhaps even for some animals and a few plants. The arguments of Julian Simon, Bjorn Lomborg, and all the other cornucopians take nourishment from the Cartesian moment and temporarily mute the critics.

The increase in knowledge and technology, especially since about 1500, the beginning of the scientific revolution and the Enlightenment, has caused us increasingly to believe in and be proud of our power of precision, our power to transform, and our predictive ability. But problems and cracks in the certainty persist. Advances in basic physics mesmerize us even as long-term weather remains difficult to anticipate, forcing forecasters to speak in probabilities. There is tacit acknowledgment of our ignorance from time to time, as when arguments are offered for saving biodiversity because in the dwindling array of life-forms are answers to questions we have not yet learned to ask.

Then there is the problem that much of our *basic* knowledge turns out to be wrong. During the 1960s, we learned that genes, meaning DNA, make messenger RNA, which makes protein for one function. By the 1990s that had changed. We have discovered that many, if not most, protein products have more than one function. In thirty short years, technology, theory expansion, and new methods combined to provide insights into the nature of life from the molecular level up.

Complex adaptive systems represent three words that combine to cover profound ignorance. We may live in civilization, but our still-wild body regulates itself minute to minute, day to day, year to year, until we die. Ignorance is still the rule.

John Maddox's *What Remains to Be Discovered* outlines four potential calamities that the author sees as awaiting us. The book is about, among other things, the future of the human race. One threat is increased levels of carbon dioxide and climate change. Another is novel and resurgent infections. Also, an asteroid could nail us. And the human genome may be unstable.[2]

Where do these threats leave us? We might be able to do something about climate change. Indeed, we seem to be doing that now. Whether it is enough to head off a catastrophe of major proportions is another question. We could use antibiotics in a more prudent way. The feedlot is the main culprit here, whether it is for cattle, hogs, chickens, or turkeys. We are not going to worry about the meteors, hoping to have a chance to duck. But instability of the human genome? Should we worry about it? Well, we don't think we should muck around with it very much. It has served us pretty well from our time in the trees, to an upright stance on the African savanna, and even through these past ten to twelve thousand years of the agricultural experiment.

Acknowledging our vast ignorance, doesn't our best opportunity to learn how to live with ignorance come from a study of nature's arrangements that have shaken down over the millennia? If so, the ecosystem becomes the conceptual tool common to all of us interested in producing food and fiber sustainably. This includes forestry and fisheries as well as agriculture since the farmer's plants and animals, the fisher's fish, and the forester's logs are all derived from ecosystems and those ecosystems have important commonalities.[3] Here we would risk an even grander extension of the ecosystem concept into other avenues of our economic life, for no other reason than the fact that a study of ecosystems invariably forces us to examine nature's economy. What that economy is all about is nothing less than efficient management and recycling of materials using contemporary sunlight.

Such an exercise is certain to force us to feature questions that go beyond the available answers. By so doing, we have taken the first step toward driving knowledge out of its categories. Where that goes, no one can predict, but it should take us in directions more inclined to give us a soft landing than the direction in which we are now headed.

We used to believe that science and technology are, ultimately, at the service of a dominating social organization. That belief now seems too simple. As Charles Washburn has said, the social-political paths available in number and design are too limited to receive the science-technology-economic outpouring.[4] What better way to deal with this reality than to begin to operate as though the twentieth century will be the last century in which we believe that knowledge is adequate to run the world? Acknowledging ignorance might be the secular mind's only way to humility. Embracing ignorance has the potential to unite the secularist with the seriously religious. By embracing an ignorance-based worldview, at least we go with our long suit.

We can't prove it, but we are betting that there is an important paradox in all this: knowledge and insight accumulate fastest in the minds of those who hold an ignorance-based worldview. Having studied the exits, their imaginations are less narrow. Darting eyes have the potential to see more.

THE CONTEXT

Matfield Green, Kansas.

This collection of essays is the product of "Toward an Ignorance-Based Worldview," one among occasional gatherings of invited guests in the Flint Hills of Kansas known as the Matfield Green Dialogues. This particular gathering, held in June 2004, was a dare of sorts. Can the modern mind undo and clear out the concepts and beliefs that wrongly led to the plowing of the prairie in the first place, to the clearing of its native and, now, most of its second-generation inhabitants, to the undoing of cultural and ecological knowing and unknowing appropriate to this place, and to the mining and exploitation of fertility, virtue, and value?

The Matfield Green Dialogues is a rare event in American letters.[5] Wes Jackson serves as host and provocateur. He provides the topics, which are designed to challenge not just common assumptions but also that common worldview with which most of us are occupied, enchanted, and paradoxically wedded to and simultaneously determined to divorce ourselves from. These meetings bring together Jackson's friends and colleagues and put them to work on big questions. In a Matfield Green classroom, without pretension or self-importance, and with the support of local folks and local food, the guests put their heads together and work it out.[6] They share meals together, take walks on the prairie, and toss and turn in the hot Kansas nights, mostly free of distractions, and with good jokes in ample supply. They have a sense, too, that the work done here might just matter a hundred years out when we're all dried up and blown away.

As his work at the Land Institute demonstrates,[7] Jackson likes to zero in on civilization's highest priorities. He has frequently remarked that, if we don't get sustainability right in agriculture, it won't matter whether we get it right anywhere else. If we can't feed our bodies without causing others to starve, or poisoning ourselves, or mining the soil, it won't make much difference whether we recycle our beer cans or use compact fluorescent lightbulbs since we are creating a future in which there isn't likely to be any work, or culture, or fertility, or prosperity, or posterity—a future, in other words, in which we cannot enjoy our beer under an electric light. Jackson and his colleagues at the Land Institute have been working for twenty-five years to "improve the security of our food and fiber source by reducing soil erosion, decreasing dependency upon petroleum and natural gas, and relieving the agriculture-related chemical contamination of our land and water." Their "specific research is an innovation for agriculture, using 'nature as the measure' to develop mixed perennial grain crops as food for humans where farmers use nature as a standard or measure in making their agronomic decisions."[8]

In the same way that agriculture underpins so much of what is vital to any culture, Jackson asked his 2004 Matfield Green guests to focus on knowledge as the primary feedstock for the Western love of progress and the scientific control of nature and to consider the possibility that, despite its flashy achievements, the knowledge-based worldview has not delivered what it promised and is getting more dangerous to practice. To put it another way, Jackson has reworked his claim about getting sustainability into agriculture to also include the mind: if a knowledge-based worldview can't feed the mind without unmaking and breaking the world, causing species extinctions and dead zones in the Gulf of Mexico, poisoning water supplies in the American Midwest, consuming without end, impoverishing the world's people with a method that was designed to set them free, creating a moral relativism that is incapable of choosing prudently between alternatives, or having the gumption to close off some alternatives altogether, then the world is better off without it, and the sooner the better. In short, knowledge alone is not adequate to run the world.

Attack the knowledge-based worldview, and its economic, scientific, political, and ethical progeny have less strength. And right now the mere promise of future knowledge freeing us from our present plights is about all the strength these other systems have going for them. If we want people to see this crumbled worldview for what it is, we have to block the only myth that makes it bearable, namely, the belief that human knowledge is sufficient to get us out of the holes we've dug for ourselves and the world. And so we are left with this: if we don't get an alternative for the current knowledge-based worldview, then the changes we make to the other systems won't matter much; they'll be more like tinkering with the engines of an airplane that has run out of fuel while in flight.

We call this view an *ignorance-based worldview,* and we predicate it on the assumption that human ignorance will always exceed and outpace human knowledge and, therefore, that before we make any decision or take any action, we must consider who and how many are involved, the level of cultural change that will be involved, and the chances of backing out if things go sour. In no way does such a view imply that we should not seek knowledge or that we are stupid or even wicked, but it does force us to remember things, cause us to hope for second chances, and provide an incentive to keep the scale small.

THE ESSAYS

The essays in this collection represent a cross section of intellectual perspectives, from engineering and agriculture to philosophy and evolutionary biology. The authors are writers, activists, physicians, natural and social scientists, philosophers, policymakers, and educators. The essays are meditative, critical, historical, and practical. There is in all of them a sense of trying to get at something new and important while also acknowledging the long history of prophets and philosophers who have advanced in nearly every generation a healthy skepticism about the reach of the human intellect and warnings about the dangers of human pride.

But this book is neither a history of these warnings nor a philosophy textbook about the many categories of knowledge and ignorance, although some of our authors do provide such categories. It is intended, rather, for readers and thinkers willing to consider a fundamental reconceptualization of our major social and cultural systems and categories; for teachers seeking to educate students for the century we are in, not the one we recently left; for students who want to challenge the status quo; and for anyone who sees the increasingly serious social and environmental problems around the globe as symptomatic of weaknesses and flaws in the very foundations of our enterprises rather than simply a matter of improper implementation of the mainstream worldview.

Which is to say: the Enlightenment worldview is problematic even when properly applied. Advances in agriculture and medicine, for example, have led to the exponential growth of the human population, and that has put increased demands on topsoil and freshwater. Technology has made more and more of the world's fossil fuels accessible, leading to increased consumption and an increase in atmospheric carbon, leading to increased global temperatures. Worse, many of the solutions to these monumental challenges depend on the logic of plenty: finding more oil, increasing soil and seed productivity, promoting economic growth and material consumption, utilizing more land for human food production, and even increasing human population. Each calls forth a faith in the unbounded human spirit to rise to any occasion, to conquer any foe. The recipe for success is simple: unleash human ingenuity; utilize it to harness and commodify nature's immense and complex forces; enjoy the new and improved world that results; repeat.

The Virtues of Ignorance challenges this prideful optimism with essays that point out the problems and dangers of a knowledge-based worldview, that outline in many styles and variations an alternative ignorance-based worldview, and that offer actual examples of such a worldview in action. Our title is intentionally provocative. We could have used terms like *prudence, humility, wisdom, precaution,* or *reverence,* all of which are implied by an ignorance-based worldview. But *ignorance* has a harsher tone and focuses attention on the very large gap between our aspirations and reality. In its traditional usage, *ignorance* has a negative connotation because it describes a human condition that is correctable. But our use of the term suggests that, no matter how much human beings discover about the natural world or ourselves or our political and social systems, knowledge will always be dwarfed by what we do not, should not, and cannot ever know. In trying to ignore or outrun ignorance, we will inevitably fail, harming others and ourselves. Ignorance of this sort is not a curable condition, and we must begin to create post-Enlightenment systems of thought that acknowledge ignorance as an initial operating condition in a living universe.

This is not to say that knowledge has no place here. Of course it does. Many of our authors are working scientists, engineers, and physicians who rely daily on traditions of knowledge to do their work, even while they create new knowledge. Robert Root-Bernstein, a scientist and one of the authors in this collection, put it well in an e-mail exchange with one of the editors:

> One can't be content with existing knowledge because any long-standing problems are long-standing precisely because existing knowledge is inadequate to address them or caused the problems in the first place. One must therefore address ignorance as well as knowledge so that we know what we need to discover or invent. But discovery and invention do not occur in a vacuum—they always build on previous knowledge. Thus, the purpose in focusing on ignorance is to build knowledge in light of what we both know and do not know. And the building of more knowledge will reveal new forms of ignorance, ad infinitum. For me, knowledge and ignorance are the yin and yang of understanding. You can't have one without the other, and when they are out of balance, the world is in trouble.

Divided into four parts, and dedicated to this balance of which Root-Bernstein speaks, the book begins with nine essays in two parts ("First Cut" and "Second Cut") that lay out the terms and represent "answers" to the question, What is an ignorance-based worldview?

Beginning "First Cut," Wes Jackson traces his interest in the ignorance worldview alternative to a letter he received from Wendell Berry in 1982 and to the Land Institute's Sunshine Farm project. Like many authors in the book, he worries about the environmental stresses and strains on a culture overly dependent on fossil fuels that some predict could be depleted in fifty years. He offers what he calls a "knowledge informed by ignorance," a perspective that admits that human "knowledge is not adequate to run this world."

Wendell Berry provides thoughtful taxonomies of both ignorance and knowledge and articulates how the ignorant application of knowledge is behind so many of our contemporary problems in science and the modern corporation. He offers advice and hope for overcoming the worst of our "arrogant ignorance."

Robert Perry offers readers a meditative discourse on information, knowledge, and wisdom. Whereas one can hope that information leads to knowledge and knowledge to wisdom, Perry suggests that our wisdom teaches us that we know little or nothing and that endless facts are not the building blocks of even the most rudimentary knowledge. He offers instead the wisdom of children and novices: to go slowly and carefully, to show compassion for others, and to embrace the world with reverence.

Richard D. Lamm claims that nothing in our past prepares us for the environmental problems that we are faced with. We cannot *grow* our way out of these problems, nor can we use history to put them in perspective. The lessons we have learned living on an empty earth are the wrong ones. We are still trying to "be fruitful, multiply, and subdue" an earth that now needs saving. Contemporary life, says Lamm, presents us with a series of problems for which the lessons of history *not only* are useless but also teach us the wrong lessons. In response to these problems, he contrasts what he calls "infinite" and "finite" cultures and suggests that our best hope lies with the latter.

Conn Nugent focuses on carbon fuels and their consequences. He offers a number of predictions and side bets about the continued consumption of these fuels and the ignorance we have about maintaining a multibillion-person population and our consumptive lifestyles without them. He challenges readers to devise support systems without depleting the common stock of natural resources.

"Second Cut" begins with the engineer Raymond H. Dean's description of how natural and cultural boundaries provide stabilizing organization in natural and cultural systems. An unbounded world would be too chaotic for any form of life. It's impossible for any one person to know everything. The illusion that we know things we don't really know frequently produces terrible mistakes. On the other hand, full awareness of our ignorance would be intolerably confusing and depressing, so, Dean says, we must learn to accept ignorance as an inevitable and even beneficial aspect of life. Dean's essay demonstrates how nature can teach us to work in a region of "optimal uncertainty" between our knowledge and our ignorance in the search for answers. In life, as in nature, failure to heed boundaries in our pursuit of knowledge leads to chaos, followed by new constraints not always to our liking. Learning to work with the constraints we have now, rather than the unexpected ones that will result when we run roughshod over the boundaries that await us, is the wiser choice.

Steve Talbott uses the metaphor of a conversation to describe an approach to understanding the world of others when perfect knowledge of this world is impossible and total ignorance unacceptable. Conversations allow us to put forth cautious questions, to make up for past inadequacies, and to acknowledge that there is never a perfect right or wrong response. Talbott encourages us to think of the art of conversation as the best approach to knowing the world around us.

Anna L. Peterson begins with two questions: First, how might ignorance provide a vantage point from which we can understand and criticize established ethical models? Second, how might we construct an ignorance-based ethics? Her essay is a survey of a whole range of knowledge- and ignorance-based secular and religious systems, with case studies of neo-Christian realism, the Reformation, and Christian existentialism. She ends her essay by speculating that an ignorance-based worldview points toward an ethical system in which action depends on possibility and hope rather than prediction.

Paul G. Heltne warns us against what he calls "purposeful ignorance," which he describes as subtle, surreptitious, and highly effective in deflecting us from achieving the humility and honesty of an ignorance-based worldview. Purposeful ignorance has us believing that we understand a situation (or at least that someone does) rather thoroughly. Purposeful ignorance often comes with an aura that its knowledge claims and practices are based on unquestionable assumptions. Thus, it would be foolish to ask questions or probe basic assumptions. Heltne believes that purposeful ignorance is deeply rooted in our cultural systems. It may seem contradictory, but the patterns of purposeful ignorance often include scientific ways of thinking and speaking. The patterns, in fact, may be pervasive. Heltne offers several illustrations of purposeful ignorance in an effort to get an initial view of what is a potent enemy. He also considers an alternative way of looking at nature that may keep us in the humility of an ignorance-based worldview.

In part 3, "Precursors and Exemplars," our authors continue to grapple with the definitions and implications of an ignorance-based worldview by providing historical background and biographical touchstones. Charles Marsh's essay opens up this part with a discussion of classical Greek philosophy and the debate over the soul's ignorance and whether this ignorance is curable. Plato's Socrates says that it is and that philosophers must impose knowledge on the common person. Aristotle claims that, in some circumstances, exact certainty is not possible, while Isocrates offers an even dimmer view of humankind's ability to know the world. This debate, claims Marsh, has appeared again two millennia later, and he encourages us to get it right this time.

For Peter G. Brown, Western society has been chasing after Francis Bacon's project of world redemption through scientific knowledge, only to be engulfed by the growing ignorance being stirred up, in part, by the project itself. With each step forward, the horizon of real progress recedes further from view. What is needed is a new worldview that orients society toward the ecosphere, toward dynamic engagement with the coevolutionary processes of the universe. Brown advances Albert Schweitzer's "reverence for life" philosophy and ethic to this end. Here we find a fulcrum for atonement, a foundation for knowledge that embraces ignorance, and the opening of a way to restore humankind to full membership in the commonwealth of life and to understand our place in the drama of the cosmos.

The philosopher Strachan Donnelley looks at the works of Alfred North Whitehead, one of the twentieth century's most important philosophers, and Ernst Mayr, the eminent evolutionary biologist. Given that Whitehead and Mayr explicitly challenge the foundation of early modern science in the name of an adequate understanding and valuation of organic, including human, life, Donnelley believes that it is time to rub together the philosophic sticks of Whitehead, Darwin, and Mayr. Specifically, Donnelley considers the contributions that Whitehead's and Darwin's modes of thought bring to the philosophic and moral exploration of the interactions of humans and nature and specifically to the discussion of the interrelations of nature, value, and democracy—to notions of ecological and democratic citizenship and our ultimate civic responsibilities to human and natural communities.

Bill Vitek's essay begins with an etymological analysis of *knowledge* and *ignorance* and settles on the phrase *joyful ignorance* to describe the state or activity of choosing to ignore the obvious, that is, ignoring what can easily or myopically be known. Joyful ignorance, he claims, is a willful, defiant, rebellious act in the face of the obvious, of certainty, of security and control, and of domination. He uses the work of Aldo Leopold to describe what he calls a more *civic way of knowing,* or the *civic mind,* and ends with some suggestions for educators and policymakers interested in promoting it.

The final part of the book, "Applications," looks at the implications of an ignorance-based worldview for public policy, education, and everyday life.

Robert Root-Bernstein shares his perspective as a working scientist who revels in saying "I don't know." He argues that what makes a scientist successful is not certainty about what we know but certainty about what we don't know. Science is not a search for solutions but a search for answerable questions. He is, consequently, disturbed by the fact that we unwittingly train science students by challenging them to answer questions to which we already have answers rather than training them to raise answerable questions that no one has ever asked. As a corrective, he provides an outline of methods that might be used to teach students how to question effectively and provides anecdotes from his own career to illustrate the fecundity of the approach he espouses.

Marlys Hearst Witte, Peter Crown, Michael Bernas, and Charles L. Witte describe the Curriculum on Medical Ignorance (CMI) that they have designed and implemented since 1984 in the University of Arizona's Department of Surgery. The curriculum is designed to help students understand the various forms of ignorance that are at work in the medical enterprise and to provide students with the appropriate navigational skills and attitudes for dealing productively with ignorance in situations where lives hang in the balance. Students learn the importance of collaboration and asking questions that go beyond the available answers and develop the values of curiosity, skepticism, humility, and optimism. And, further, they come to realize that, as they gain more knowledge, exponentially more questions are uncovered, providing the fuel for learning, discovery, and medical advances.

The economist Herb Thompson takes neoclassical economics to task—particularly as it is expressed in the classroom—for actively promoting what he calls "ignorance-squared," or the purposeful uncoupling of the economy from its messy sociocultural and political contexts, and the use of mathematical models and theoretical constructs (the market, the rational consumer, cost-benefit analysis) to blind students to what the discipline of economics doesn't know. The curriculum is filled with too many concepts and not enough time for reflection, the cultivation of perspective, the presentation of opposing viewpoints, or the knowledge and ignorance produced by other disciplines. It is a state of affairs not unlike that prevailing in many other disciplines, but it is particularly troubling and dangerous in economics. Thompson suggests some alternative pedagogical strategies that give the student of economics a healthy and productive dose of ignorance.

Jon Jensen explores some possible ignorance-based worldview answers to three basic questions: What is education for? How does education prepare people for the work of the world within a culture and a place? What sort of person are we striving to create? Jensen admits that this notion of educating for ignorance is a counterintuitive idea and one that is vulnerable to the dangerous misinterpretation that we should not value education. But he suggests that the idea is equally ripe in its prospects for helping us rethink education in a way that is essential to the health of our culture and the biosphere. Jensen gives ignorance meaning in an educational context and makes a number of suggestions for education reform that might help us produce a new type of graduate, with a new vision for the future.

Joe Marocco puts an ignorance-based worldview to the test by applying its principles to climate change. Against the backdrop of Francis Bacon's own words about the nature of knowledge, Marocco demonstrates that the central problem with climate change is environmentalists' insistence that it is a problem to be solved the old-fashioned way rather than a complex and interactive set of problems and relationships requiring many perspectives and techniques. He takes knowledge-based environmentalism to task and offers a broad and ignorance-based alternative to grappling with climate change.

The book ends with an essay by Craig Holdrege, a teacher and scientist. Holdrege invites us to resist the many abstractions we have come to take for granted and to see instead with fresh eyes. By stripping away the many layers that separate us as subjective perceivers from the living and natural world that we both perceive and interact with, our vision and our relationships to what we see are restored. Holdrege's observations and personal experiences provide the reader with a way back to a living, unpredictable world.

THE UPSHOT

Collectively, the essays in this book take seriously the charge that the knowledge-based worldview that has governed so much of Western culture for half a millennium is both flawed and dangerous. The reader will find analyses of current social and environmental challenges, criticisms of the knowledge-based perspective that has brought us to this place, definitions and perspectives, alternatives and optimism. What comes next is, admittedly, harder to discern, but all the authors in this collection go beyond mere critique of our knowledge-based scientific, economic, and technological paradigms and actually offer concepts and insights into what a worldview based on fundamental human ignorance and limitations would look like. To put it another way: What would human cultures look like, and how might we interact differently in the world, if we began every endeavor and conversation with the humbling assumption that human understanding is limited by an ignorance that no amount of additional information can mitigate? How would we educate our children differently or engage in scientific research? Might we be more cautious and more willing to listen to others—and not just other human beings, but the whole conversation going on all around us? The essays in this collection effectively begin that conversation.

Whether it is sustainability, civic science, the precautionary principle, ecological economics, natural systems agriculture, industrial ecology, or any number of other new approaches and concepts, there is a growing consensus that the Western world needs to rethink many of its major axioms. The value of this collection is that it goes to the common core of this axiomatic trouble: a belief as old as the Genesis story of Adam and Eve stealing the knowledge of good and evil, and made manifest in the Greek claim that reason makes humans the measure of all things, and refined still in the European Enlightenment, where human liberty is, finally, unbounded by wholesale revolutions in politics, science, technology, and economics. It is a belief that ultimately trounces and marginalizes the natural world, indigenous cultures, and, generally, the more soft-spoken and humble alternatives: community, craft, place, folk knowledge, oral traditions, and thrift. Fortunately, many who have mastered and benefited most from this belief system are starting to get roughed up as well. If this belief has run its course and is now perceived to be more dangerous than useful, more hubris than honest, and simply unable to deliver what it had promised by way of wealth, freedom, and the domination of nature, then shouldn't we be asking what comes next?

We believe that this collection of essays represents a serious and thought-provoking discussion, not only of what is misguided about our current knowledge-based worldview, but also of what might replace it. We hope that these essays provoke awareness and some deep thinking about how "the still unlovely human mind"—Aldo Leopold's words—conceives the world and how we might improve our minds first and foremost while giving the world a respite from our attempts to remake it.

NOTES

1. We refer here to Robert Costanza et al., "The Value of the World's Ecosystem Services and Natural Capital," *Nature* 387, no. 6630 (15 May 1997): 253–60; and Gretchen Daily, *Nature's Services: Societal Dependence on Natural Ecosystems* (Washington, DC: Island, 1997).

Objections have been raised that these eminent men and women were operating with the same mind-set as the promoters of our current extractive economy. We sympathize with the objectors, but there is the fact that, especially during the past quarter century, more and more people have insisted that ecosystem services, like rain for irrigation or freezes that kill pests, and the value of human capital be counted on all corporate and government balance sheets.

2. John Maddox, *What Remains to Be Discovered: Mapping the Secrets of the Universe, the Origins of Life, and the Future of the Human Race* (New York: Martin Kessler, 1998).

3. We realize that more than ecosystems are at issue here. For example, humans will never, for all practical purposes, completely understand the nervous system, let alone the nature of all the feedback loops and intersecting variables within the rest of the human body. It is too complicated.

4. Charles Washburn is a friend of Jackson's.

5. The first of the Matfield Green Dialogues was called "Toward a Taxonomy of Boundaries," the second was the subject of this book, and the third will be titled "The Ecosphere as Real and Conceptual Tool."

6. The classroom is in the Matfield Green public school, which was closed in 1973 and donated to the Land Institute in 1993 to be used as a cultural and educational center.

7. For more information, see http://www.landinstitute.org.

8. The "History" from which the quotes in the text are taken can be found under the link "About Us" on the Land Institute's Web site (see n. 7 above).

PART ONE

First Cut

TOWARD AN IGNORANCE-BASED WORLDVIEW

Wes Jackson

My inspiration for considering a worldview based on ignorance came from two sources. The earliest was the following letter from Wendell Berry:

> Port Royal, Kentucky
> July 15, 1982
>
> Dear Wes,
>   I want to try to complete the thought about "randomness" that I was working on when we talked the other day.
>   The Hans Jenny paragraph that started me off is the last on page 21 of *The Soil Resource* [Jenny 1980]:
>
> > "Raindrops that pass in random fashion through an imaginary plane above the forest canopy are intercepted by leaves and twigs and channeled into distinctive vert space patterns of through-drip, crown-drip, and stem flow. The soil surface, as receiver, transmits the 'rain message' downward, but as the subsoils lack a power source to mold a flow design, the water tends to leave the ecosystem as it entered it, in randomized fashion."
>
> My question is: Does "random" in this (or any) context describe a verifiable condition or a limit of perception?
>   My answer is: It describes a limit of perception. This is, of course, not a scientist's answer, but it may be that *anybody's*

answer would be unscientific. My answer is based on the belief that pattern is verifiable by limited information, whereas the information required to verify randomness is unlimited. As I think you said when we talked, what is perceived as random within a given limit may be seen as part of a pattern within a wider limit.

If this is so, then Dr. Jenny, for accuracy's sake, should have said that rainwater moves from mystery through pattern back into mystery.

If "mystery" is a necessary (that is, honest) term in such a description, then the modern scientific program has not altered the ancient perception of the human condition a jot. If, in using the word "random," scientists only mean "random so far as we can tell," then we are back at about the Book of Job. Some truth meets the eye; some does not. We are up against mystery. To call this mystery "randomness" or "chance" or a "fluke" is to take charge of it on behalf of those who do not respect pattern. To call the unknown "random" is to plant the flag by which to colonize and exploit the unknown. (A result that our friend Dr. Jenny, of course, did not propose and would not condone.)

To call the unknown by its right name, "mystery," is to suggest that we had better respect the possibility of a larger, unseen pattern that can be damaged or destroyed and, with it, the smaller patterns.

This respecting of mystery obviously has something or other to do with religion, and we moderns have defended ourselves against it by turning it over to religion specialists, who take advantage of our indifference by claiming to know a lot about it.

What impresses me about it, however, is the insistent practicality implicit in it. If we are up against mystery, then we dare act only on the most modest assumptions. The modern scientific program has held that we must act on the basis of knowledge, which, because its effects are so manifestly large, we have assumed to be ample. But if we are up against mystery, then knowledge is relatively small, and the ancient program is the right one: Act on the basis of ignorance. Acting on the basis of ignorance, paradoxically, requires one to know things, remember things—for instance, that failure is possible, that error is possible, that second chances are desirable (so don't risk everything on the first chance), and so on.

What I think you and I and a few others are working on is

a definition of agriculture as up against mystery and ignorance-based. I think we think that this is its *necessary* definition, just as I think we think that several kinds of ruin are the *necessary* result of an agriculture defined as knowledge-based and up against randomness. Such an agriculture conforms exactly to what the ancient program, or programs, understood as evil or hubris. Both the Greeks and the Hebrews told us to watch out for humans who assume that *they* make all the patterns.

Berry's explanation of the practical value of ignorance came in the midst of the Land Institute's Sunshine Farm project, a ten-year study directed by the late Marty Bender. This project was designed to determine what percentage of the food the farm produced would be available for export if the farm ran exclusively on contemporary sunlight collected by crops on the farm and photovoltaic panels. To be fair, we needed to estimate the embodied energy of both living and nonliving arrangements on the 210 acres, including tractors, photovoltaic panels, draft horses, chickens, bolts, fencing, medicine for the cattle herd, and on and on. A major problem Marty had was determining where to draw the line for energy expenditures off the farm. We realized that we can't know with any high degree of certainty. We're ignorant.

It was hard to find practical value with this sort of ignorance. After all, if, when analyzing a problem, we could agree where to draw a line both in time and in space, countless arguments could be avoided. Some of us think that with the oil peak we'll be at the moment in history when the slack cut by abundance will start to fade, finally forcing us to account for our energy use, which is causing rapid climate change. Then, as accountants, we will become students of boundaries. To do the bookkeeping historically has required degrees of democratic agreement and dictatorial imperative. We have understood this as necessary in households and in government. But the boundary of consideration has been narrower than the boundary of causation. Deficit spending of ecological capital has been the rule and is nearly always given short shrift, perhaps because we don't want to face its difficulty.

Embedded within civilization is scaffolding that we take for granted but that must be considered if we want to know the embodied energy of any human artifact and what it takes to repair it. And when we think of what gave us the slack to make the new arrangements called the *advancement of civilization,* when we think of what stands behind the supply of basic needs such as clothing and shelter, surplus food, and all the geegaws we want or think we need, we realize that no accounting sheet would ever have enough space.

This problem made me wonder: Was there ever a time since our gathering and hunting days that the planet's capital stock has not been drawn down to support agriculture and civilization? Imagine viewing a one-hour film of our ten-thousand-year journey since agriculture began. Imagine our food supply and what energy-rich carbon stands behind our crops. The film would show that, over those ten millennia, we have glided from one energy-rich carbon pool to another. First it was soil carbon, then forest carbon, then coal, then oil, then natural gas. The film would show that, in the wake of agricultural expansion, the potential of most agricultural landscapes to fix carbon *without subsidy from elsewhere* has declined. Ability and desire to track these costs would have given us entirely different concepts of efficiency and progress. Instead, they are masks, painted by large-group fanaticism. In light of our profound ignorance, they become either comic or sources of derision.

This all did lead to something. It is a hypothesis: *Since agriculture began, humans have produced no technological product or process—including our crops and livestock—without drawing down the earth's capital stock and, thereby, reducing the overall net primary production of its ecosystems using only contemporary sunlight. In shorthand: We cannot do better than nature.*

Certainly by the time cells arrived, and maybe even before, life on the earth, like all other events in the known universe, has had to operate within the constraints of the entropy law, the second law of thermodynamics. For some, this is like stating that gravity exists. But we need reminding because often in our accounting we overlook what the great physicist Max Planck saw: "I found the meaning of the Second Law of Thermodynamics in the principle that in every natural process the sum of the entropies of all bodies involved in the process increases" (1968, 18). I am not a physicist or a physical scientist, but in numerous discussions with physicists I have been comforted to learn that many of them acknowledged that, in the whole field of physics, entropy is the concept that is the most difficult to understand and, at the same time, the most fundamental.

If we think of Planck's understanding, the hypothesis about drawdown carries a corollary: Darwinian selection pressure operates all through the structural hierarchy of cells, tissues, organs, organisms, ecosystems, and ecosphere. All things considered, optimum or nearly optimum efficiencies are achieved through integration. In such a manner, it seems to me, is "the sum of the entropies" minimized. Selection pressure is on all organisms in the system, on infrastructure we have yet to comprehend, and, more important, on infrastructure we will never comprehend—and, thankfully, need not understand. Acknowledging our ignorance, we recognize the need to protect the integrity of wild ecosystems that have experienced minimal human disruption. These ecosystems become our standards because they represent the best example of what is optimally efficient anywhere across the ecological mosaic of the ecosphere. (Note that I did not call it the *biosphere,* which places an organism or bio bias on the subject.) Natural ecosystems generally have higher levels of net primary production than do agricultural systems, even with farming's fossil carbon subsidies (Field 2001).

The hypothesis is probably not falsifiable and, therefore, violates Popper's assertion of the necessity of such to be valid. Count it as having heuristic knowledge-based value.

The usual argument in favor of technological solutions, which depend on what we know, is that whether we can do better than nature depends on the specific technology. The argument goes something like this: "Sure, humans build bulldozers and bombs, but they have also created energy-efficient windows." Energy-efficient the windows may be, but the total underlying structure, the scaffolding, including bulldozers and bombs, contributed to those windows. The infrastructure supporting windowhood coming out of civilization itself came at the cost of the dismemberment of efficient integrations somewhere, likely many places, within the ecosphere. There is no way we can account for the precise cost to the ecosphere, even at one moment, or for the time it will take it to repair itself or, if it can be repaired, to reach the level of former net primary production. Twigs and leaves falling from trees to the forest floor, and eventually trunks and limbs, will break down faster than an equal weight of iron. But for the forest or the prairie it is a short loop back to renewal. Contrast that with the long loops and energy cost to assemble iron atoms from ore, make them useful, and protect them from rust or return them to the junkyard to be recycled again. Contemporary sunlight favors the forest ecosystem. Because the economy of nature works without us, without our knowledge of every minute detail, we can be comfortable in our ignorance. But the technological infrastructure for making an iron beam comes at the cost of nicking away at the ecosphere. Recovery is likely, but the self-repair of the ecosphere is slower than civilization's assault.

This hypothesis is dismal but, if taken seriously, is useful in our time, when technology assessment must move to a central position in our culture. That rapid climate change is upon us is widely accepted now. We watch the ascent of the population curve and the incomprehensible difficulties sure to come now that we approach the oil peak in the near distance. If technology assessment is based on the assumption that we humans can do better than nature in our creative, technological, knowledge-based efforts, meaning that we will not draw down the earth's capital stock, the criteria for decisionmaking will be strikingly different from the opposite assumption, the one of the hypothesis. Let's imagine that Party A's core philosophy is that we can do better than nature. Party B says that we cannot do better than nature. Party C is agnostics who say: "Sometimes we can, and sometimes we cannot." Assume that the full technological array, the full tool kit, is available to all three parties. Parties A and B have assumptions based on faith. Party C relies on the idea that knowledge is adequate to know when we can and when we cannot build and use a gadget without a drawdown of the earth's capital.

To this point in the discussion I have described our attempts to estimate the embodied energy of an object or a technology and how we realized we were on an ignorance trail. It is not complete ignorance, however. We can list many ways that our ancestors and we have spent the earth's capital: eroded soils, energy-rich soil carbon, clear-cut forests that cannot return to a former level, trees turned into charcoal to smelt iron and build ships, coal for the industrial revolution, oil, natural gas. On this ten-thousand-year journey, we see the five pools of energy-rich carbon extracted from, burned, or wasted, ecosystems dismembered. Ignorant about the details we may be, but there is a bottom line: the net primary production of nature's ecosystems has been reduced. Our ability to determine the cost of any particular technological product may go to zero, but the aggregate of the cost to the earth's sequestered resources and the services of the ecosphere and its ecosystems stands as a stark reminder of a way of life dependent on subtraction since agriculture began.

Wendell Berry once said to me that we have "never known what we are doing because we have never known what we have been undoing." As we slowly come to appreciate our embeddedness within the ecosphere, we become mindful that what sustains us is not our dazzling industrial output but what we don't very well understand. Looking back on our journey of knowledge accumulation in the midst of blissful ignorance, we finally realize that the context for our existence is better defined by an understanding of ecology, rather than of industry. The primary source of destruction in our time is industrial.

When we examine the context of destruction, we realize, as Wendell Berry says in his essay in this volume, that few of the proposed solutions are "commensurate with our problems" and that the "great problems call for many small solutions." When we lose small farms and rural communities, we lose cultural information, cultural capacity, and know-how for more solar-based livelihoods, livelihoods more evenly distributed over the ecological mosaic. To rely primarily on industrial solutions to correct ecological damage is to call on the same mind. Here enters the discipline of ecology. The physical mosaic the earth presents, minerals in the soil, precipitation, temperature ranges, winds, all those things, as well as all the organisms within these ecosystems, make up an aggregate that has featured material recycling running on contemporary sunlight—for aeons. Given that we have far more ignorance than knowledge about the workings within, the ecosystem becomes the best conceptual tool to help us understand how to get along in this world. Its study forces us see the abiotic and the biotic as one. Other levels in the structure hierarchy—organisms, organs, tissues, cells, molecules, atoms—won't do. It is this larger slab of space/time in which we live, the ecosystem, where most of the accounting has already been done and where our ignorance of the ecosphere and its accounting systems will always exceed our knowledge.

To reiterate, at the risk of belaboring the point, none of this is to argue against a conventional understanding of energy-efficient technology. Bring on the energy-efficient windows. I am simply arguing that even the most efficient technologies draw down the capital that powers ecosystems to renewably maximize carbon fixation. Ecosystems tend to accumulate ecological capital. Civilizations spend it, and the depreciation rate is huge. Yet most of what makes those windows and other cultural artifacts is never figured in. That is what makes all these "efficient" technologies suspect. Whether we are building photovoltaic cells or wind machines, flat plate collectors or nuclear power plants, the standing infrastructure that made them possible is mostly sponsored by drawdown of one or more of the five carbon pools.[1]

The geographer Carl Sauer once noted that the industrial revolution was made possible by plowing the world's temperate grasslands (see Sauer 1981, 357). The grains of the newly opened grasslands' fertility were exported to emerging industrial centers, mostly in Western Europe, at very low prices, with no reckoning of the maintenance of the soil sponsoring that abundance, let alone the forests going into the ships for both timber and the charcoal necessary to make iron. Either we were woefully ignorant about this reality, or we didn't care. It was probably both.

Ecology and evolutionary biology were immature disciplines when that plowing was well under way in America. Darwin's *On the Origin of Species* was published in 1859.[2] The "entangled bank" of interconnected species that Darwin describes toward the end of that great work was what we now call an ecosystem. Darwin began as a geologist, but he also admired the geographer Alexander von Humboldt. He was well equipped to advance his ideas of habitat, integrating the biological, physical, and chemical. The farmers of those grasslands, the merchants of their grain, and nearly all the rest were not, as that integration was plowed asunder. What mattered to them was that the industrial revolution gained from the released fertility, the accumulated interest from the journeywork of ecosystems past. Then coal, oil, and natural gas covered for falling natural fertility to prop up agricultural output. Public notice was scant. A sustainable management of nutrients, organic matter, and water was crippled.

THE ROLE OF GEOLOGIC TIME AS A PRIMARY SOURCE OF SUSTENANCE AND HEALTH

There are twenty-some nutrients in our soils that go into life. Plants need these nutrients to capture from the atmospheric commons life's other four requisite elements—carbon, hydrogen, oxygen, and nitrogen. A combination of wind, water, gravity, and geologic activity such as mountain formation makes the land-based elements available. The Upper Midwest's soils came mostly from another activity also measured in geologic time, 1.7-million-year Ice Age glaciation.

I emphasize geologic time to make the elemental source available because the challenge for nature's ecosystems is to keep those twenty-some soil-based elements within reach and manage them efficiently. The land has to be protected from above with cover and from below with a net of life, a diversity of roots. We may be ignorant of the precise details of how it all works, but we know in general that species diversity both above and below is a source of dependable chemistry. In the seldom-seen teeming diversity below, bacteria, fungi, and invertebrates reproduce by the power of sun-sponsored photons captured in the green molecular traps set above. If we could adjust our eyes to a power greater than the electron microscope, our minds would reel in a seemingly surrealistic universe of ions exchanging where water molecules dominate and colloidal clay plates are held by organic thread molecules important in a larger purpose but regarded as just another meal by innumerable microscopic invertebrates. As roots decay and aboveground residues break down, released nutrients tumble through soil catacombs to start all over again. And we who stand above in thoughtful examination, smelling and rolling fresh dirt between our fingers, distill these myriad actions into one concept—soil health—and leave it at that.

When historians speak of a civilization going into decline, little attention is paid to the decline in the natural fertility of the soils that sustained that civilization. Egypt has the largest long-lasting civilization. This should be no surprise, given the quantity of nutrients, from mountains forming in Ethiopia, that go down the Blue Nile to meet the White Nile carrying organic matter from central Africa's steamy jungles. We might say that flood-borne nutrients built the pyramids, using humans as agents. But it is more. To be totally ridiculous, the inventory of events behind those pyramids could carry us back to the Big Bang and quantum physics. But it is worth remembering that at each step and intersection on that journey are aggregates of activity we can never fathom.

We don't have to go back to the Big Bang measured in billions of years, but what deserves to be emphasized again and again is that the mineral recharge that increases carbon-fixing potential usually happens in *geologic* time, millions of years, not agricultural time, 10,000 years, or industrial time, 250 years. The beginning of agricultural time closely matches the beginning of the end of the Stone Age. Geologic activity through geologic time built up much of the bank account for humanity's credit card appropriated 10,000 years ago. As gatherer-hunters we more or less lived within our means. Once we were on our way on the agricultural journey, natural fertility deposits made at geologic speed seldom kept pace with our wasting of the soil resource, which amounts to wasteful spending of the bank account. This is an old problem long acknowledged. To correct the wasteful use, individuals and societies have tried to imitate nature in their practices. Hugh Hammond Bennett, the founding chief (in 1935) of the U.S. Soil Conservation Service (SCS), said in a 1958 interview: "We tried to imitate nature. We abided by the following basic physical facts: (1) land varies greatly from place to place, due to differences in soil, slope, climate and vegetative adaptability; (2) land must be treated according to its natural capability and its condition as the result of the way man has used it; (3) slope, soil and climate largely determine what is suitable protection in all situations. But above all, we tried to imitate nature" (Martin 1958, 48).

Beyond these words, Bennett did not elaborate on what he meant when he said, "We tried to imitate nature." In his time, he had such intellectual allies as Liberty Hyde Bailey, J. Russell Smith, and Sir Albert Howard. All were clear about the necessity to learn more from nature. As Wendell Berry has written (see Berry 1990), they were contemporary descendants of a stream of intellectual ancestors in the literary tradition and elsewhere. They had all articulated the necessity and advocated processes that were tried.

Unfortunately, a cursory review of the past sixty years of agricultural history will show that the knowledge-as-adequate worldview became the primary assumption of industrial agriculture. The efforts of Bennett's people and essentially all who followed were sidetracked. Fossil fuels made it possible. Fossil fuels made it compelling since they were tied to an economy becoming dependent at an exponential rate. Fossil fuel–based fertility was able to offset the decline of natural fertility. It may seem that, with more fossil fuel going into agriculture, good practices should have been easier. That is asking too much for most places over time precisely because humans are poor micromanagers of nutrients and water. As humans, we are forced to operate at the macro scale. Moreover, when it comes to grain production, at least in recent time, neither the USDA nor the land grant system set out to act on the obvious belowground reality of most of nature's land-based "nature," the perennial root mixture, what the Land Institute soil scientist Jerry Glover calls the elegant "micro-management in millimeters and minutes" of nutrients and water. As a consequence, our predecessors in Bennett's time, and before, were promoting an agriculture that cannot be sustained. It cannot be sustained because the primary high-yielding crops, which command most of the acreage, feature annual roots and monocultures. Annual crops grown in strips or planted on the contour will help some in reducing erosion, but not enough. Multiple annual species grown in strips is not a polyculture in the sense that a prairie or a forest is a polyculture. Nearly a half century after the interview with the SCS founding chief, we still have erosion. We have added serious chemical contamination of our groundwater and land and a dead zone in the Gulf of Mexico (one of 150 certified around the world) all since the green revolution techniques, based on knowledge, have been put into practice. Neither knowledge nor practice is adequate to stem the destruction. Given the reality of annual roots, adequate management of the problem is beyond our ethical stretch. Bennett, like his contemporaries, lacked the means to meaningfully implement the "nature as measure" ideal. Had perennial roots on the major crops been available and been planted in mixtures that mimic the vegetative structure of native prairie, conservation could have been—for the first time in the history of grain production—a consequence of production. As it stands now, and as it has stood in the past, if annual roots are all that is available, essentially all conservation is the consequence of subtraction. And so we have the current destructive institutional dualism exemplified in the USDA today with the Natural Resource Conservation Service devoted to conservation and the Agriculture Research Center devoted to production.

Given the complexity of an ecosystem, it seems ironic that two changes like perennials and polycultures can make such a difference. The diversity in the architecture of perennial roots provides the essential network for dealing with the reality of the physical diversity with which life has to contend. Root penetration is really an interpenetration into the world of which we are made and yet about which we are largely ignorant. It is the world that determines our future possibilities. Why it has been so largely ignored is a big question. It is dark down there, dark and, to some, dirty, and so the entire assembly is regarded as dirt. Even if we could see that world on a daily basis, it would still be "through a glass darkly," for we will never understand that complexity that sustains us, but neither do we have to. Let ignorance in this realm reign.

THE EXERCISE OF KNOWLEDGE INFORMED BY IGNORANCE

I belong to Party B—we can never do better than nature. Nevertheless, we can do better than we are currently doing. For the first time in human history, we can build those roots into several of our crops. As recently as Bennett's time, scientists lacked the knowledge to make wide crosses between annuals and perennials. Embryo rescue did not exist then. Scientists at the Land Institute use it as a matter of routine in one of our small, inexpensive labs. A knowledge of genomes and several insights into genetics that we now take for granted were not part of my genetics training in the mid-1960s. Understanding the genome will help in a general way in the future. We can use molecular markers now. We can "paint" chromosomes if we need or want to. And, of course, Bennett's era lacked the computational power we now take for granted.

Bennett did not mention polycultures. If we are to build domestic grain-producing prairies, we will need to assemble working polycultures. We don't have to wait on ecology to mature. That discipline is sufficiently mature to be useful now if we had the perennials up and running. Published results in the twin disciplines of ecology and evolutionary biology have been getting bought and paid for over the past eighty years and more. Countless studies sit on the shelf, largely untapped, ready to be employed as needed.

Many questions still lie ahead. We don't know how many plant species we will need for our domestic grain-producing prairie to function effectively. We do have suggestive data gathered over many years. How much redundancy such systems will need is unknown. The most important functional components (warm and cool season grasses, legumes, sunflowers) can only be guessed at now. We know that it is the interactions that are responsible for health and stability. Interactions are across lateral as well as vertical gradients, both above and below the surface. And, though I believe that all these technological advances and insights are dependent on the scaffolding of civilization, they can be used to slow down the nicking of the ecosphere.

It is hard to be optimistic. We may add as many people to the planet in the next fifty years as the planet had a half century ago. In the last half century, which included the green revolution, featuring heavy fossil fuel subsidies allowed us to more than double the food supply. To double that increase once more seems likely to include encroachment on more wild land. The human "footprint," the result mostly of agriculture, is causing what is increasingly called a *biodiversity crisis*. The prospects for protecting the earth's biodiversity sound bleak. Market incentives are not aligned to protect the wild. Returns to agricultural capital are marginal. Perverse subsidies have led to overproduction and land clearing. Funding for conservation is limited. Producers increasingly do not own the land, so the incentive to conserve is minimal. There is a lack of clear land tenure in the developing world. Increasingly, large producers, agribusiness interests, which include the lenders and commodities groups, call the shots. Extensive clearing of the rain forest, where most of the biodiversity lies, and intensive approaches to agriculture are occurring simultaneously. Consolidation is accelerating. Governments are corrupt. Research is focused on a simplification model producing single commodities rather than multiple benefits. That farmers do not like what is going on suggests that it is out of control. All these problems are the product of working knowledge rather than an acknowledgment of vast ignorance.

How do we protect the wildness we need as a standard against which to judge our agricultural practices? Ecologists with little interest in agriculture may be inclined to propose that we intensify our inputs on the agricultural lands in order to avoid encroachment on the wild biodiversity. Ecology has taught us that the planet has no seams, that all is interconnected both in time and in space. Cores of Greenland ice record the Iron Age of the Roman Empire. The dead zone in the Gulf of Mexico has its origin eleven hundred miles away. Agriculture is a major contributor to the demise of fisheries. Intensifying agriculture via the industrial model in order to save biodiversity is likely to have the opposite effect.

An effective strategy to preserve biodiversity must encompass both wilderness sanctuaries and working agricultural landscapes. Sanctuaries are critical for maintaining gene pools. But they cannot enclose migratory species, and they lie vulnerable to invasions from abutting farmlands. An exotic crop can displace indigenous plants. A government can reduce nature preserves in order to provide more arable land for a hungry population, as happened in India following the departure of the British. For biodiversity sanctuaries to work, it is necessary to have farms that are both (*a*) compatible with local ecosystems and (*b*) sufficiently productive and resource efficient to feed large populations over a long time.

At the Land Institute, we believe that a transition to a biodiversity-friendly agriculture could be accomplished globally within fifty years. The first releases for farms featuring perennial grain mixtures in North America could come sooner. If we had a fifty-year farm bill along with the five-year variety, key milestone objectives could be measured at these five-year intervals. Ecologists and agriculturalists will have to be joined at the hip to make the transition work. There is no reason why ecologists, who have had the luxury of being descriptive, and agriculturists, who have had the burden of being prescriptive, cannot meet. They have to because the seriousness of the separation of these two cultures is far greater than the seriousness of the separation of the two cultures described by C. P. Snow (1993).

We live in a fallen world relying on the tree of knowledge. In a limited sense, it has been a "great run" since these first seeds were planted ten millennia ago. Without those seeds and the soils into which they were planted, there would have been no pyramids, no Parthenon, no temple of Solomon, no Teotihuacan, no Forbidden City, no Chartres, no San Francisco, no United States, no Kansas. There would have been no Job, no Aristotle, no Virgil, no Dante, no Shakespeare. And, without the later subsidy of fossil fuels, in combination with our soils and our grains, the scientific revolution would have stalled. There would be no Einstein equations, no knowledge of DNA, no Hubble telescope, no knowledge of tectonic plates and continental drift, no knowledge of geologic or cosmic time, no expansion of our knowledge of the scale of the universe or the inner recesses of the atom. All are part of the "great run" of civilization, which required soil and seeds.

Cave paintings done by preagriculturists are wonderful, and probably the uncodified oral tradition was too, but, beyond all that, we are the only species in this part of the universe that knows that we are made of stardust recycled through supernova at least twice. Awareness of our stellar origins should be a sign that we are capable of absorbing the big lessons of our planet's ecosystems and then applying those lessons to agriculture. The agriculture we seek will act like an ecosystem, feature material recycling, and run on the contemporary sunlight of our star. By acting on our goal to make agriculture sustainable, we will have taken the first step forward for humanity to measure its progress by its independence from the extractive economy. Will we ever become totally independent of such an economy? Probably not until civilization is gone. But we can do better; we can do a lot better. We can slow it all down, and maybe buy another ten thousand years.

T. S. Eliot's four lines from "Little Gidding" (1942) provide us with the most hopeful outcome:

We shall not cease from exploration
And the end of all our exploring
Will be to arrive where we started
And know the place for the first time. (Eliot 1980, 145)

We will not know the place for the first time unless we admit and act on our profound ignorance. This requires that we admit that knowledge is not adequate to run this world. The challenge is to be able to act on that admission when conventional thinking has a way of reasserting itself.

NOTES

1. Nutrient deposits in the Mediterranean, the Gulf of Mexico, and elsewhere, shaken loose by the plow and pulled there by gravity, would have to be returned to the distant hillsides by the likes of wind machines and nuclear reactors to keep our food supply at a more or less constant level.

2. I can't resist inserting here that, in the same year, one Colonel Drake drilled the first oil well, in Pennsylvania. Darwin's insight about evolution through natural selection had been partly sponsored by coal and England's forests—they let the British rule the waves following the defeat of the Spanish Armada. Darwin's insight was probably not possible without deficit spending of the capital stock that sustained the context in which he lived. Most modern scientists' refinements of his insight since have been sponsored by the slack afforded by oil and natural gas.

REFERENCES

Bailey, Liberty Hyde. 1915. *The Holy Earth.* New York: Scribner's.

Berry, Wendell. 1990. "A Practical Harmony." In *What Are People For?* 103–8. San Francisco: North Point.

———. 1997. *Home Economics.* San Francisco: North Point.

Cohen, Joel E. 1995. *How Many People Can the Earth Support?* New York: Norton.

Eliot, T. S. 1980. *The Complete Poems and Plays, 1909–1950.* New York: Harcourt, Brace.

Field, Christopher B. 2001. "Sharing the Garden." *Science* 294, no. 5551 (21 December): 2490–91.

Hong, Sungmin, Jean-Pierre Candelone, Clair C. Paterson, and Claude F. Boutron. 1994. "Greenland Ice Evidence of Hemispheric Lead Pollution Two Millennia Ago by Greek and Roman Civilizations." *Science* 265, no. 5180 (23 September): 1841–43.

Jenny, Hans. 1980. *The Soil Resource: Origin and Behavior.* New York: Springer.

Hillel, Daniel J. 1991. *Out of the Earth: Civilization and the Life of the Soil.* New York: Free Press.

Howard, Sir Albert. 1956. *An Agricultural Testament.* 7th impression. London.

Martin, Santford. 1958. "And History Is Already Shining on Him—Some Impressions of Hugh H. Bennett." In *Better Crops with Plant Food,* 43–55. Washington, DC: American Potash Institute.

Miller, G. Tyler, Jr. 1971. *Energetics, Kinetics, and Life: An Ecological Approach.* Belmont, CA: Wadsworth.

Planck, Max. 1968. *Scientific Autobiography, and Other Papers.* Translated by Frank Gaynor. New York: Greenwood. Originally published in 1949.

Rabalais, Nancy N., R. Eugene Turner, and Donald Scavia. 2002. "Beyond Science into Policy: Gulf of Mexico Hypoxia and the Mississippi River." *BioScience* 52, no. 2 (February): 129–42.

Sauer, Carl O. 1981. *Selected Essays, 1963–1975.* Berkeley, CA: Turtle Island Foundation.

Smith, J. Russell. 1950. *Tree Crops.* New York: Harper & Row.

Snow, C. P. 1993. *The Two Cultures.* Cambridge: Cambridge University Press. Originally published in 1959.

THE WAY OF IGNORANCE

Wendell Berry

In order to arrive at what you do not know
You must go by a way which is the way of ignorance.
—T. S. Eliot, "East Coker"

Our purpose here is to worry about the predominance of the supposition, in a time of great technological power, that humans either know enough already, or can learn enough soon enough, to foresee and forestall any bad consequences of their use of that power. This supposition is typified by Richard Dawkins's assertion, in an open letter to the Prince of Wales, that "our brains . . . are big enough to see into the future and plot long-term consequences."

When we consider how often and how recently our most advanced experts have been wrong about the future, and how often the future has shown up sooner than expected with bad news about our past, Mr. Dawkins's assessment of our ability to know is revealed as a superstition of the most primitive sort. We recognize it also as our old friend hubris, ungodly ignorance disguised as godly arrogance. Ignorance plus arrogance plus greed sponsors "better living with chemistry," and produces the ozone hole and the dead zone in the Gulf of Mexico. A modern science (chemistry or nuclear physics or molecular biology) "applied" by ignorant arrogance resembles much too closely an automobile being driven by a six-year-old or a loaded pistol in the hands of a monkey. Arrogant ignorance promotes a global economy while ignoring the global exchange of pests and diseases that must inevitably accompany it. Arrogant ignorance makes war without a thought of peace.

We identify arrogant ignorance by its willingness to work on too big a scale, and thus to put too much at risk. It fails to foresee bad consequences not only because some of the consequences of all acts are inherently unforeseeable, but also because the arrogantly ignorant often are blinded by money invested; they cannot afford to foresee bad consequences.

Except to the arrogantly ignorant, ignorance is not a simple subject. It is perhaps as difficult for ignorance to be aware of itself as it is for awareness to be aware of itself. One can hardly begin to think about ignorance without seeing that it is available in several varieties, and so I will offer a brief taxonomy.

There is, to begin with, the kind of ignorance we may consider to be inherent. This is ignorance of all that we cannot know because of the kind of mind we have—which, I will note in passing, is neither a computer nor exclusively a brain, and which certainly is not omniscient. We cannot, for example, know the whole of which we and our minds are parts. The English poet and critic Kathleen Raine wrote that "we cannot imagine how the world might appear if we did not possess the groundwork of knowledge which we do possess; nor can we in the nature of things imagine how reality would appear in the light of knowledge which we do not possess."

A part of our inherent ignorance, and surely a most formidable encumbrance to those who presume to know the future, is our ignorance of the past. We know almost nothing of our history as it was actually lived. We know little of the lives even of our parents. We have forgotten almost everything that has happened to ourselves. The easy assumption that we have remembered the most important people and events and have preserved the most valuable evidence is immediately trumped by our inability to know what we have forgotten.

There are several other kinds of ignorance that are not inherent in our nature but come instead from weaknesses of character. Paramount among these is the willful ignorance that refuses to honor as knowledge anything not subject to empirical proof. We could just as well call it materialist ignorance. This ignorance rejects useful knowledge such as traditions of imagination and religion, and so it comes across as narrow-mindedness. We have the materialist culture that afflicts us now because a world exclusively material is the kind of world most readily used and abused by the kind of mind the materialists think they have. To this kind of mind, there is no longer a legitimate wonder. Wonder has been replaced by a research agenda, which is still a world away from demonstrating the impropriety of wonder. The materialist conservationists need to tell us how a materialist culture can justify its contempt and destructiveness of material goods.

A related kind of ignorance, also self-induced, is moral ignorance, the invariable excuse of which is objectivity. One of the purposes of objectivity, in practice, is to avoid coming to a moral conclusion. Objectivity, considered a mark of great learning and the highest enlightenment, loves to identify itself by such pronouncements as the following: "You may be right, but on the other hand so may your opponent," or "Everything is relative," or "Whatever is happening is inevitable," or "Let me be the devil's advocate." (The part of devil's advocate is surely one of the most sought after in all the precincts of the modern intellect. Anywhere you go to speak in defense of something worthwhile, you are apt to encounter a smiling savant writhing in the estrus of objectivity: "Let me play the devil's advocate for a moment." As if the devil's point of view will not otherwise be adequately represented.)

There is also ignorance as false confidence, or polymathic ignorance. This is the ignorance of people who know "all about" history or its "long-term consequences" in the future. And this is closely akin to self-righteous ignorance, which is the failure to know oneself. Ignorance of one's self and confident knowledge of the past and future often are the same thing.

Fearful ignorance is the opposite of confident ignorance. People keep themselves ignorant for fear of the strange or the different or the unknown, for fear of disproof or of unpleasant or tragic knowledge, for fear of stirring up suspicion and opposition, or for fear of fear itself. A good example is the United States Department of Agriculture's panic-stricken monopoly of inadequate meat inspections. And there is the related ignorance that comes from laziness, which is the fear of effort and difficulty. Learning often is not fun, and this is well-known to all the ignorant except for a few "educators."

And finally there are for-profit ignorance, which is maintained by withholding knowledge, as in advertising, and for-power ignorance, which is maintained by government secrecy and public lies.

Kinds of ignorance (and there must be more than I have named) may thus be sorted out. But having sorted them out, one must scramble them back together again by acknowledging that all of them can be at work in the same mind at the same time, and in my opinion they frequently are.

I may be talking too much at large here, but I am going to say that a list of kinds of ignorance comprises half a description of a human mind. The other half, then, would be supplied by a list of kinds of knowledge.

At the head of that list let us put the empirical or provable knowledge of the materialists. This is the knowledge of dead certainty or dead facts, some of which at least are undoubtedly valuable, undoubtedly useful, but at best this is static, smallish knowledge that always is what it always was, and it is rather dull. A fact may thrill us once, but not twice. Once available, it is easy game; we might call it sitting-duck knowledge. This knowledge becomes interesting again when it enters experience by way of use.

And so, as second, let us put knowledge as experience. This is useful knowledge, but it involves uncertainty and risk. How do you know if it is going to rain, or when an animal is going to bolt or attack? Because the event has not yet happened, there is no empirical answer; you may not have time to calculate the statistical probability even on the fastest computer. You will have to rely on experience, which will increase your chance of being right. But then you also may be wrong.

The experience of many people over a long time is traditional knowledge. This is the common knowledge of a culture, which it seems few of us any longer have. To have a culture, mostly the same people have to live mostly in the same place for a long time. Traditional knowledge is knowledge that has been remembered or recorded, handed down, pondered, corrected, practiced, and refined over a long time.

A related kind of knowledge is made available by the religious traditions and is not otherwise available. If you premise the falsehood of such knowledge, as the materialists do, then of course you don't have it, and your opinion of it is worthless.

There also are kinds of knowledge that seem to be more strictly inward. Instinct is inborn knowledge: how to suck, bite, and swallow; how to run away from danger instead of toward it. And perhaps the prepositions refer to knowledge that is more or less instinctive: *up, down, in, out,* etc.

Intuition is knowledge as recognition, a way of knowing without proof. We know the truth of the Book of Job by intuition.

What we call conscience is knowledge of the difference between right and wrong. Whether or not this is learned, most people have it, and they appear to get it early. Some of the worst malefactors and hypocrites have it in full; how else could they fake it so well? But we should remember that some worthy people have believed conscience to be innate, an "inner light."

Inspiration, I believe, is another kind of knowledge or way of knowing, though I don't know how this could be proved. One can say in support only that poets such as Homer, Dante, and Milton seriously believed in it, and that people do at times surpass themselves, performing better than all you know of them has led you to expect. Imagination, in the highest sense, is inspiration. Gifts arrive from sources that cannot be empirically located.

Sympathy gives us an intimate knowledge of other people and other creatures that can come in no other way. So does affection. The knowledge that comes by sympathy and affection is little noticed—the materialists, I assume, are unable to notice it—but in my opinion it cannot be overvalued.

Everybody who has done physical work or danced or played a game of skill is aware of the difference between knowing how and being able. This difference I would call bodily knowledge.

And finally, to be safe, we had better recognize that there is such a thing as counterfeit knowledge or plausible falsehood.

I would say that these taxonomies of mine are more or less reasonable; I certainly would not claim that they are scientific. My only assured claim is that any consideration of ignorance and knowledge ought to be at least as complex as this attempt of mine. We are a complex species—organisms surely, but also living souls—who are involved in a life-or-death negotiation, even more complex, with our earthly circumstances, which are complex beyond our ability to guess, let alone know. In dealing with those circumstances, in trying "to see into the future and plot long-term consequences," the human mind is neither capacious enough nor exact nor dependable. We are encumbered by an inherent ignorance perhaps not significantly reducible, as well as by proclivities to ignorance of other kinds, and our ways of knowing, though impressive within human limits, have the power to lead us beyond our limits, beyond foresight and precaution, and out of control.

What I have said so far characterizes the personal minds of individual humans. But because of a certain kind of arrogant ignorance, and because of the gigantic scale of work permitted and even required by powerful technologies, we are not safe in dealing merely with personal or human minds. We are obliged to deal also with a kind of mind that I will call corporate, although it is also political and institutional. This is a mind that is compound and abstract, materialist, reductionist, greedy, and radically utilitarian. Assuming as some of us sometimes do that two heads are better than one, it ought to be axiomatic that the corporate mind is better than any personal mind, but it can in fact be much worse—not least in its apparently limitless ability to cause problems that it cannot solve, and that may be unsolvable. The corporate mind is remarkably narrow. It claims to utilize only empirical knowledge—the preferred term is "sound science," reducible ultimately to the "bottom line" of profit or power—and because this rules out any explicit recourse to experience or tradition or any kind of inward knowledge such as conscience, this mind is readily susceptible to every kind of ignorance and is perhaps naturally predisposed to counterfeit knowledge. It comes to its work equipped with factual knowledge and perhaps also with knowledge skillfully counterfeited, but without recourse to any of those knowledges that enable us to deal appropriately with mystery or with human limits. It has no humbling knowledge. The corporate mind is arrogantly ignorant by definition.

Ignorance, arrogance, narrowness of mind, incomplete knowledge, and counterfeit knowledge are of concern to us because they are dangerous; they cause destruction. When united with great power, they cause great destruction. They have caused far too much destruction already, too often of irreplaceable things. Now, reasonably enough, we are asking if it is possible, if it is even thinkable, that the destruction can be stopped. To some people's surprise, we are again backed up against the fact that knowledge is not in any simple way good. We have often been a destructive species, we are more destructive now than we have ever been, and this, in perfect accordance with ancient warnings, is because of our ignorant and arrogant use of knowledge.

Before going further, we had better ask what it is that we humans need to know. We need to know many things, of course, and many kinds of things. But let us be merely practical for the time being and say that we need to know who we are, where we are, and what we must do to live. These questions do not refer to discreet categories of knowledge. We are not likely to be able to answer one of them without answering the other two. And all three must be well answered before we can answer well a further practical question that is now pressing urgently upon us: How can we work without doing irreparable damage to the world and its creatures, including ourselves? Or: How can we live without destroying the sources of our life?

These questions are perfectly honorable, we may even say that they are perfectly obvious, and yet we have much cause to believe that the corporate mind never asks any of them. It does not care who it is, for it is not anybody; it is a mind perfectly disembodied. It does not care where it is so long as its present location yields a greater advantage than any other. It will do anything at all that is necessary, not merely to live, but to aggrandize itself. And it charges its damages indifferently to the public, to nature, and to the future.

The corporate mind at work overthrows all the virtues of the personal mind, or it throws them out of account. The corporate mind knows no affection, no desire that is not greedy, no local or personal loyalty, no sympathy or reverence or gratitude, no temperance or thrift or self-restraint. It does not observe the first responsibility of intelligence, which is to know when you don't know or when you are being unintelligent. Try to imagine an official standing up in the high councils of a global corporation or a great public institution to say, "We have grown too big," or "We now have more power than we can responsibly use," or "We must treat our employees as our neighbors," or "We must count ourselves as members of this community," or "We must preserve the ecological integrity of our workplaces," or "Let us do unto others as we would have them to do unto us"—and you will see what I mean.

The corporate mind, on the contrary, justifies and encourages the personal mind in its worst faults and weaknesses, such as greed and servility, and frees it of any need to worry about long-term consequences. For these reliefs, nowadays, the corporate mind is apt to express noisily its gratitude to God.

But now I must hasten to acknowledge that there are some corporations that do not simply incorporate what I am calling the corporate mind. Whether the number of these is increasing or not, I don't know. These organizations, I believe, tend to have hometowns and to count themselves as participants in the local economy and as members of the local community.

I would not apply to science any stricture that I would not apply to the arts, but science now calls for special attention because it has contributed so largely to modern abuses of the natural world, and because of its enormous prestige. Our concern here has to do immediately with the complacency of many scientists. It cannot be denied that science, in its inevitable applications, has given unprecedented extremes of scale to the technologies of land use, manufacturing, and war, and to their bad effects. One response to the manifest implication of science in certain kinds of destruction is to say that we need more science, or more and better science. I am inclined to honor this proposition, if I am allowed to add that we also need more than science.

But I am not at all inclined to honor the proposition that "science is self-correcting" when it implies that science is thus made somehow "safe." Science is no more safe than any other kind of knowledge. And especially it is not safe in the context of its gigantic applications by the corporate mind. Nor is it safe in the context of its own progressivist optimism. The idea, common enough among the universities and their ideological progeny, that one's work, whatever it is, will be beneficently disposed by the market or the hidden hand or evolution or some other obscure force is an example of counterfeit knowledge.

The obvious immediate question is, How *soon* can science correct itself? Can it correct itself soon enough to prevent or correct the real damage of its errors? The answer is that it cannot correct itself soon enough. Scientists who have made a plausible "breakthrough" hasten to tell the world, including of course the corporations. And while science may have corrected itself, it is not necessarily able to correct its results or its influence.

We must grant of course that science in its laboratories may be well under control. Scientists in laboratories did not cause the ozone hole or the hypoxic zones or acid rain or Chernobyl or Bhopal or Love Canal. It is when knowledge is corporatized, commercialized, and applied that it goes out of control. Can science, then, make itself responsible by issuing appropriate warnings with its knowledge? No, because the users are under no obligation to heed or respect the warning. If the knowledge is conformable to the needs of profit or power, the warning will be ignored, as we know. We are not excused by the doctrine of scientific self-correction from worrying about the influence of science on the corporate mind, and about the influence of the corporate mind on the minds of consumers and users. Humans in general have got to worry about the origins of the permission we have given ourselves to do large-scale damage. That permission is our problem, for by it we have made our ignorance arrogant and given it immeasurable power to do harm. We are killing our world on the theory that it was never alive but is only an accidental concatenation of materials and mechanical processes. We are killing one another and ourselves on the same theory. If life has no standing as mystery or miracle or gift, then what signifies the difference between it and death?

To state the problem more practically, we can say that the ignorant use of knowledge allows power to override the question of scale, because it overrides respect for the integrity of local ecosystems, which respect alone can determine the appropriate scale of human work. Without propriety of scale, and the acceptance of limits which that implies, there can be no form—and here we reunite science and art. We live and prosper by form, which is the power of creatures and artifacts to be made whole within their proper limits. Without formal restraints, power necessarily becomes inordinate and destructive. This is why the poet David Jones wrote in the midst of World War II that "man as artist hungers and thirsts after form." Inordinate size has of itself the power to exclude much knowledge.

What can we do? Anybody who goes on so long about a problem is rightly expected to have something to say about a solution. One is expected to "end on a positive note," and I mean to do that. But I also mean to be careful. The question, What can we do? especially when the problem is large, implies the expectation of a large solution.

I have no large solution to offer. There is, as maybe we all have noticed, a conspicuous shortage of large-scale corrections for problems that have large-scale causes. Our damages to watersheds and ecosystems will have to be corrected one farm, one forest, one acre at a time. The aftermath of a bombing has to be dealt with one corpse, one wound at a time. And so the first temptation to avoid is the call for some sort of revolution. To imagine that destructive power might be made harmless by gathering enough power to destroy it is of course perfectly futile. William Butler Yeats said as much in his poem "The Great Day":

Hurrah for revolution and more cannon shot!
A beggar upon horseback lashes a beggar on foot.
Hurrah for revolution and cannon come again!
The beggars have changed places, but the lash goes on.

Arrogance cannot be cured by greater arrogance, or ignorance by greater ignorance. To counter the ignorant use of knowledge and power we have, I am afraid, only a proper humility, and this is laughable. But it is only partly laughable. In his political pastoral "Build Soil," as if responding to Yeats, Robert Frost has one of his rustics say,

> I bid you to a one-man revolution—
> The only revolution that is coming.

If we find the consequences of our arrogant ignorance to be humbling, and we are humbled, then we have at hand the first fact of hope: We can change ourselves. We, each of us severally, can remove our minds from the corporate ignorance and arrogance that is leading the world to destruction; we can honestly confront our ignorance and our need; we can take guidance from the knowledge we most authentically possess, from experience, from tradition, and from the inward promptings of affection, conscience, decency, compassion, even inspiration.

This change can be called by several names—*change of heart, rebirth, metanoia, enlightenment*—and it belongs, I think, to all the religions, but I like the practical way it is defined in the Confucian *Great Digest*. This is from Ezra Pound's translation:

> The men of old wanting to clarify and diffuse throughout the empire that light which comes from looking straight into the heart and then acting, first set up good government in their own states; wanting good government in their states, they first established order in their own families; wanting order in the home, they first disciplined themselves; desiring self-discipline, they rectified their own hearts; and wanting to rectify their hearts, they sought precise verbal definitions of their inarticulate thoughts [the tones given off by the heart]; wishing to attain precise verbal definitions, they set to extend their knowledge to the utmost.

This curriculum does not rule out science—it does not rule out knowledge of any kind—but it begins with the recognition of ignorance and of need, of being in a bad situation.

If the ability to change oneself is the first fact of hope, then the second surely must be an honest assessment of the badness of our situation. Our situation is extremely bad, as I have said, and optimism cannot either improve it or make it look better. But there is hope in seeing it as it is. And here I need to quote Kathleen Raine again. This is a passage written in the aftermath of World War II, and she is thinking of T. S. Eliot's poem *The Waste Land,* written in the aftermath of World War I. In *The Waste Land,* Eliot bears unflinching witness to the disease of our time: We are living the death of our culture and our world. The poem's ruling metaphor is that of a waterless land perishing for rain, an image that becomes more poignant as we pump down the aquifers and dry up or pollute the rivers:

> But Eliot [Kathleen Raine said] has shown us what the world is very apt to forget, that the statement of a terrible truth has a kind of healing power. In his stern vision of the hell that lies about us . . . , there is a quality of grave consolation. In his statement of the worst, Eliot has always implied the whole extent of the reality of which that worst is only one part.

Honesty is good, then, not just because it is a virtue but for a practical reason: It can give us an accurate description of our problem, and it can set the problem precisely in its context.

Honesty, of course, is not a solution. As I have already said, I don't think there are solutions commensurate with our problems. I think the great problems call for many small solutions. But for that possibility to attain sufficient standing among us, we need not only to put the problems in context but also to learn to put our work in context. And here is where we turn back from our ambitions to consult both the local ecosystem and the cultural instructions conveyed to us by religion and the arts. All the arts and sciences need to be made answerable to standards higher than those of any art or science. Scientists and artists must understand that they can honor their gifts and fulfill their obligations only by living and working as human beings and community members rather than as specialists. What this may involve may not be predictable even by scientists. But the best advice may have been given by Hippocrates: "As to diseases make a habit of two things—to help, or at least, to do no harm."

The wish to help, especially if it is profitable to do so, may be in human nature, and everybody wants to be a hero. To help, or to try to help, requires only knowledge; one needs to know promising remedies and how to apply them. But to do no harm involves a whole culture, and a culture very different from industrialism. It involves, at the minimum, compassion and humility and caution. The person who will undertake to help without doing harm is going to be a person of some complexity, not easily pleased, probably not a hero, probably not a billionaire.

The corporate approach to agriculture or manufacturing or medicine or war increasingly undertakes to help at the risk of harm, sometimes of great harm. And once the risk of harm is appraised as "acceptable," the result often is absurdity: We destroy a village in order to save it; we destroy freedom in order to save it; we destroy the world in order to live in it.

The apostles of the corporate mind say, with a large implicit compliment to themselves, that you cannot succeed without risking failure. And they allude to such examples as that of the Wright brothers. They don't see that the issue of risk raises directly the issue of scale. Risk, like everything else, has an appropriate scale. By propriety of scale we limit the possible damages of the risks we take. If we cannot control scale so as to limit the effects, then we should not take the risk. From this, it is clear that some risks simply should not be taken. Some experiments should not be made. If a Wright brother wishes to risk failure, then he observes a fundamental decency in risking it alone. If the Wright airplane had crashed into a house and killed a child, the corporate mind, considering the future profitability of aviation, would count that an "acceptable" risk and loss. One can only reply that the corporate mind does not have the householder's or the parent's point of view.

I am aware that invoking personal decency, personal humility, as the solution to a vast risk taken on our behalf by corporate industrialism is not going to suit everybody. Some will find it an insult to their sense of proportion, others to their sense of drama. I am offended by it myself, and I wish I could do better. But having looked about, I have been unable to convince myself that there is a better solution or one that has a better chance of working.

I am trying to follow what T. S. Eliot called "the way of ignorance," for I think that is the way that is appropriate for the ignorant. I think Eliot meant us to understand that the way of ignorance is the way recommended by all the great teachers. It was certainly the way recommended by Confucius, for who but the ignorant would set out to extend their knowledge to the utmost? Who but the knowingly ignorant would know there is an "utmost" to knowledge?

But we take the way of ignorance also as a courtesy toward reality. Eliot wrote in "East Coker":

The knowledge imposes a pattern, and falsifies,
For the pattern is new in every moment
And every moment is a new and shocking
Valuation of all we have been.

This certainly describes the ignorance inherent in the human condition, an ignorance we justly feel as tragic. But it also is a way of acknowledging the uniqueness of every individual creature, deserving respect, and the uniqueness of every moment, deserving wonder. Life in time involves a great freshness that is falsified by what we already know.

And of course the way of ignorance is the way of faith. If enough of us will accept "the wisdom of humility," giving due honor to the ever-renewing pattern, accepting each moment's "new and shocking / Valuation of all we have been," then the corporate mind as we now have it will be shaken, and it will cease to exist as its members dissent and withdraw from it.

IGNORANCE—AN INNER PERSPECTIVE

Robert Perry

Let's pause for a moment and think about what we, as human beings, need to contend with regarding our profound ignorance of how the natural world, in its most expansive sense, works.

At the collective level, an immense amount of information has accrued, particularly over the last century. However, it has done little to diminish humanity's overall ignorance, largely because the information has generally occurred in the form of countless new bits, billions of facts that are not well integrated into a larger framework of understanding. Even so, if we consider the depth of human culture and history, we do discover a form of integrated information that has been handed down over millennia: cultural knowledge. And, looking more deeply, if we examine cultural knowledge and note those elements that have helped humanity not only to survive but to improve itself, we discover a third level of highly integrated knowledge: wisdom.

I have used three different terms: *information, knowledge,* and *wisdom.* We might hope that an accumulation of information leads to knowledge and that an inspired use of knowledge leads to wisdom. This does occur. A distinct refining takes place at each step when one moves from unintegrated information to knowledge to wisdom. Once some level of wisdom has been attained, it can be used to guide the use of knowledge and even the kind of information that should be developed, accessed, or accumulated to direct the formation of further knowledge.

I say all this to show that human beings are not absolutely and completely ignorant. However, we are pretty close to that state. Today's wise men and women tell us that we know nearly nothing, that our knowledge is profoundly incomplete, and that the billions of facts and bits of information we have gathered and have at our disposal are nearly inconsequential when compared to what might be understood if we knew all the whats, wherefores, and whys of the universe.

FIRST MEDITATION: ESSENTIALS OF BEHAVIOR

Astonishing ignorance is our shared realm. And, because we dwell in this arena of darkness, we note that there are categories of ignorance. For example, there is ignorance of facts—as seen in ignorance of science, of languages, of tropical life-forms, of anatomy, and of types of stars. There's also ignorance of how to relate: emotional ignorance. In addition, there is spiritual ignorance. But what is meant by *spiritual?* I mean a grasp—or perhaps just an awareness—of the whys and wherefores of the universe. It seems that the most successful societies have always submitted somewhat gracefully to their ignorance, primarily because they were able to invent stories, myths, and explanations for the great variety of things about which they basically knew nothing. However, some societies learned that that which helped them to succeed over the long term was the ability to embrace the other—to connect with other human beings, other life-forms, with the so-called inanimate world, and with the wider cosmos. It is the ability to love: to expand the personal being so that it encompasses an ever-widening sphere of other beings.

How are human beings to behave in light of their colossal ignorance? To answer this question, we should examine what a person does when he or she is not sure, for example, of how to move in an unlit room that harbors many children sleeping on the floor. One moves, of course, with great care. In light of humanity's immense lack of integrated understanding, there is, fortunately, a way to live that encourages long-term survival and well-being: with enormous sensitivity. Perhaps that is why many of the great religions, and in particular their mystical offshoots, talk about a caring, inclusive way of relating to everything, because—given our extraordinary ignorance—that's the only way to succeed: to move forward mindful that much can be damaged if we simply lurch about in that room with children sleeping on the floor.

Observe how a parent holds his newborn child. He knows nothing about that child's physiology—what the little one is feeling, thinking, all the things that are working together physiologically to create and sustain that small being in his arms. Behaving with exceptional sensitivity and tenderness is what most clearly makes sense in the face of colossal ignorance about the infant.

If humanity ever finally comes to understand everything completely, it might conclude that people must behave with great care and tenderness. That conclusion would be virtually the same counsel that spiritual guides have been offering for millennia. In other words, an ignorance-based worldview would lead us to behave in exactly the same way as a full-knowledge worldview would. Human beings must be exquisitely aware of one another (and everything else) as much as possible.

SECOND MEDITATION: RECOMMENDED GENERAL BEHAVIOR

Maybe it makes some sense to take a closer look at how we might want to behave, given our stunning ignorance. This is a particularly good moment to consider such a thing, given the global changes presently taking place as a result of our stumbling around in ignorance.

Even so, we notice that countless generations—billions of people—acting in and through their ignorance over millennia, somehow managed to survive. Those few among them who possessed wisdom saw that they were operating within a great sphere of ignorance. Consequently, they acted (and urged others to act) with sensitivity, humility, with caution, and, most often, with reverence, understanding that, if they assumed too much or thought of themselves as the ultimate wielders of power because of what they thought they knew, they would inevitably overstep their bounds and invite misfortune, if not outright calamity. Humanity finds itself in that position right now, having experienced multiple disasters in the past century owing to enhanced, but very partial, knowledge: knowledge without the accompanying humility and reverence needed to interact with one another and whole-planet ecological systems.

Perhaps one of our biggest problems is that we forget from time to time just how little we know, and then we have to suffer the consequences. At present, however, it is not just humanity that suffers the adverse outcomes of its ignorance but virtually all forms of life on the earth.

Let us use an example of how we might choose to behave in light of our colossal ignorance. Children, being curious like most human beings, sense that they know little. Still, they forge ahead in the great chain of experiments that we call growing up. And, in that process, they learn a great deal. By the way, our efforts to learn should never stop at either the individual or the collective level, for those efforts are never in vain. Childhood, adolescence, and much of early adulthood constitutes an extended period of trial-and-error learning. Science is something like that too—but more formalized and layered with many rules. For those whose insight tells them that science is nothing more than child's play, they are correct. And, if they say so in the right spirit, they demonstrate wisdom.

Children, then, seeing that they know little, do several things: they remain curious and eager to know, so they experiment and test, an approach that reflects an appropriately cautious, sensitive method of acquiring knowledge. Sometimes this method results in injury, as when a child puts her finger on that hot stove burner "to see what will happen." This also occurs with adults, but with far less caution, with a great deal more hubris and arrogance. We set off atomic bombs to see what will happen. Will they ignite the atmosphere? Will they penetrate the planet's crust? Metaphorically speaking, these things did happen. Everyone saw just how much pure energy was packed into matter in order to create what we think of as "solid" reality. This understanding continues to make many pause, as it helped reveal a fundamental understanding: that nothing really exists here in our physical reality save energy, albeit packaged in a wide variety of forms. Humanity put its hand on the hot stove burner to see what would happen and got badly burned. For a short while after, it tried to proceed with greater caution. But now the stove tops have multiplied. They are all around us. And that proliferation is occurring in part because people have forgotten—or simply never understood—the effect of uncontrolled conversion of "matter" into energy. We are moving forward blinded by our astonishing ignorance.

That caution of the child at the stove top, and of humanity with atomic energy, is appropriate and necessary, but it is insufficient given the pace at which our ignorance is bringing about ecological chaos. Biologists especially have become intensely aware of the destruction of life-forms under way around the planet. An additional ingredient is needed, then, in the mix of behaviors needed to cope with our ignorance: respect, and perhaps even a deep affection, for that which might be lost if humanity's experimenting (or, worse, its sense of sufficient knowledge) becomes too predominant. This is why the ethos of experimental procedure has evolved and, perhaps, why ethics in general has developed with such great depth and intricacy. When people begin to assume that they know the "right" thing to do, they are not infrequently wrong, so we have gradually cultivated laws to moderate individual and collective behavior, particularly when given the fact that people generally operate within a sphere of immense ignorance. At present, this broader understanding is fueling a move toward wider recognition and adoption of a "precautionary principle" in the conduct of human affairs. It is an excellent idea.

Humanity, then, should modify its behavior in the following ways, given its vast ignorance:

- Admit its great lack of knowledge and understanding with regard to nearly everything and slow down.
- Encourage curiosity and its individual/institutional expressions (trial-and-error learning, science, religion) as long as any conclusions drawn are understood to be tentative. That is, while it is true that knowledge builds on previous knowledge, the original or fundamental premises are often revised. Consequently, any conclusions drawn from them should have the status of opinion rather than fact. The enterprise of science, at its heart, embraces this understanding. Religions would do well to follow suit.
- Understand that humanity is part of a far greater whole—that we are not separate from the rest of the cosmos. How humanity behaves, therefore, matters greatly. Consequently, it should proceed sensitively, with humility, respect, and affection regarding the planet on which it resides, including all its inhabitants.
- Educate itself (especially its children) about what it does not know in addition to what it purports to know. Most of what humanity thinks it understands now will change dramatically over the next few millennia. This may help diminish humanity's arrogance and situate its accomplishments within a far larger, integrated framework.

THIRD MEDITATION: THE PARTICULARS OF INDIVIDUAL BEHAVIOR

The only proper context in which to consider the human ignorance/behavior question is within the sphere of wisdom. A few souls walking the planet possess great wisdom. They genuinely know why things exist; they have a sense of the very long-term circumstance of human and nonhuman existence. However, their evolution and inner work is not (or, at least, not easily) transferable to others. They are, however, able to greatly inspire. Their minds are not filled with innumerable facts, bits of data, or pieces of information. Instead, their neurology is preoccupied with behavior—their own and that of others. They are concerned with how to integrate their understandings with their actions, that is, how to behave, how to be, given their grasp of a much larger framework. By observing them, we can garner useful lessons.

First, creativity is extremely important to them. Therefore, they spend considerable time creating—not just in creating art in the classical sense (music, dance, sculpture, architecture, painting), but in landscaping their interior with magnificent ideas, thoughts, and aspirations that apply to humanity as a whole. Because they understand that they are inseparable from the rest of humanity (indeed, from everything that exists), their ideas and behavior are expressed rarely for themselves alone and most often on behalf of humanity.

Second, such individuals strive to disappear. This means that they don't care about being noticed, about "making a splash," or about advancing a personal agenda. This calm demeanor invariably means that people barely notice them, though they are generally well thought of for their efficiency, collegiality, dependability, attentiveness, and helpfulness. They are present to get the job done, and then they move on to the next necessary task.

Third, they do not waste time. Nothing is unimportant to them. Most are likely to grant equal importance to, say, weeding a patch of carrots and helping deliver a baby. They are extraordinarily focused, even while remaining calm. And, though possessed of great equanimity, they are capable of quick, decisive action. Each moment is spent doing something fully and well, even if it is resting or observing an ant crawl over the face of a peony flower.

Fourth, animals, plants, soil, and stone are extremely important to them. Animals are considered younger brothers. Plants are seen to be the lungs of the world. All are part of the great "current" of living forms that constitute the distinctive fingerprint of the earth. Soil, mammals, herbs, insects—all are revered.

Fifth, they are very concerned about what children are exposed to, experience, learn, perceive, and offer. For them, children are the eyes and thoughts and actions of the future. If children's early experiences are shabby, are not filled with great love and caring, and are not informed by a sense of awe and expansion, then challenge and limitations will again appear in our collective future.

And, sixth, such individuals live simply. They minimize wants, satisfying essential needs so that they might continue acting with alertness and efficiency. Consequently, such souls are generally not plagued by discomforts, fears, depression, and mood swings. Laughter, however, comes easily to them, though it is a genuine expression of delight and nothing else.

How, then, does a wise human being—one who completely embraces the immensity of his or her ignorance—behave? With care, sensitivity, simplicity, curiosity, calm, focus, good will, and reverence for all that exists.

HUMAN IGNORANCE AND THE LIMITED USE OF HISTORY

Richard D. Lamm

Let me state up front my thesis. I believe that history has become of significantly reduced usefulness for human wisdom and for guidance in the management of the future. I believe that many of the great and wise sayings concerning the importance of history—like Santayana's that "those who cannot remember the past are condemned to repeat it" (Santayana 1905, 284) or Harry Truman's to the effect that the only surprises in the future are the history you don't know—while still true for human events, do not give us guidance on our major environmental public policy challenges and can be downright dangerous as we face the next generation of public issues. In some ways, history has become a trap because it prevents us from recognizing the full seriousness of the new problems we are faced with. An old world is dying, and a new world in which history is of limited use is struggling to be born. Heretical words, but let me make my case.

History *does* teach us much about human nature, about human ambition, cruelty, folly, about the seduction of power, the temptation of riches and lust. We enlarge our knowledge and enrich our soul by the study of history.

But history does not teach us about Mother Nature; it does not allow us to properly evaluate something like global warming, environmental degradation, or the growth of human numbers. It is likely that we are entering a new era of sustainability. A study of history would not have predicted the Renaissance or the industrial revolution, and I don't believe it is of much help in the search for the new world of sustainability or what may exist on the backside of the Hubbert curve.

The past gives little guidance to the next generation of problems because (*a*) we are living on the upper shoulder of some unprecedented and dangerous geometric curves and (*b*) we face an ecological crisis beyond historic precedent.

We ignore Al Bartlett's wise words that the greatest human failure is our inability to understand the exponential function (Bartlett, n.d.). Events are moving at stunning speed. Every modern year is the equivalent of decades of historic years. Our eighty-mile-per-hour cars need better headlights than did our parents' forty-mile-per-hour cars, but the speed of change still makes the landscape a blur. We have difficulty focusing on the patterns and what they might have taught us in more placid times. We are awash in disorienting, discontinuous change.

But, more important, the next sequences of geometric growth in human numbers and environmental impact are, I believe, unsustainable and thus, by definition, without precedent. The relentless cascade of ecological challenges is giving us a world where historical precedent is less useful. History teaches us of human limitations but not of nature's limits. History gives us little guide to a world that needs to turn from "growth" to sustainability. Some guidance, of course, but not where it really counts. We are sailing on uncharted waters.

I believe that we are surrounded with evidence that increasingly shows that something is fundamentally wrong with our historic ways of looking at the world. Yesterday's solutions have become today's problems, and these problems are of a different scale and coming at us with increasing velocity. The growth paradigm that allowed us to create wealth, reduce poverty, and increase living standards is becoming obsolete. Those human traits that allowed us to prevail over the ice, the tiger, and the bear—in a time of an empty earth—continue to operate long after we are no longer an empty earth.

In *The Spirit in the Gene,* Reg Morrison suggests that those genes that saved us as a species now are on course to destroy us (1999, 257). We are hardwired by survival traits that now, unless controlled, will drive us into oblivion. Evolution moves too slowly to correct the dilemma that evolution put us in by its past slow progress.

Our globe is warming, our forests are shrinking, our water tables are falling, our ice caps are melting, our coral is dying, and our fisheries are collapsing. Our soils are eroding, our wetlands are disappearing, our deserts are encroaching, and our finite water is more and more in demand. I suspect these to be the early warning signs of a world approaching its carrying capacity. We cannot call on the lessons of history to help us evaluate the seriousness of these problems because it is an entirely new paradigm. Ecologically, we are sailing on uncharted waters while moving at unprecedented speed. We have lost our anchor, and our navigational instruments are out of date. We are going to a whole new destination.

Nothing in our past prepares us for the environmental problems that we are faced with. We cannot *grow* our way out of these problems; both population growth and economic growth are likely to be problems rather than solutions. Living on an empty earth has taught us the wrong lessons. We are still trying to "be fruitful, multiply, and subdue" an earth that now needs saving. Contemporary life is a rock rolling downhill, gathering speed. It presents us with a series of problems of nature, for which history often teaches us the wrong lessons.

Kenneth Boulding said that the modern human dilemma is that all our experience deals with the past yet all our problems are challenges of the future (see Boulding 1974). The lessons we have learned in the past do not help and in many ways are counterproductive in solving the problems of sustainability. We cannot grow our way out of growth-created problems. Our standard economic models have become ecologically unsustainable. We face the real possibility of Easter Island on a global scale. Alas, the isolated exception has possibly become the general rule; it is a whole new world where new and different values, customs, traditions, and lifestyles will be required.

Humans appear throughout history to be insatiable creatures. There appears at this time to be no reasonable limit on "more," or "bigger," or "faster," or "richer." If we haven't already hit carrying capacity, it is just a matter of time.

The well-known Rio Declaration on Environment and Development states unequivocally: "Human beings are at the centre of concerns for sustainable development. They are entitled to a healthy and productive life in harmony with nature" (United Nations Conference on Environment and Development 1992, principle 1). These words speak to my heart but not to my head. It seems to me that in a limited ecological world we cannot entitle everyone to a healthy and productive life. Humans must fit within ecology. It seems too much to hope for that our aspirations and the aspirations of our children and grandchildren will somehow magically fit within ecological limits. I recognize that I might be wrong; we are all fragile human beings who confront the world from our own set of eyes. But, here, I have been essentially paraphrasing the United Nations, the National Academy of Sciences, and the Royal Society, all of which have warned that increasing population and increasing consumption threaten to overshoot the earth's ecological carrying capacity (see Royal Society and National Academy of Sciences 1997). Most of these historic ways that societies have grown and developed may be obsolete, and, if they are, we are overdriving our headlights and heading for major traumas.

We cannot be growth maximizers and ecological realists at the same time. The Rio Declaration was beautifully written, and I hope that what it envisions is possible. But it seems to me that you cannot maximize all the variables in any equation, and, thus, it seems improbable that we can realistically ensure that humanity meets the needs of the present without compromising the ability of future generations to meet their own needs.

We cannot solve growth-related problems with more growth; we must move to sustainability. It took a billion years or more for nature to create the limited stocks of petroleum and mineral wealth that modern technology and human ingenuity have recently learned to exploit. But we are squandering our onetime inheritance of cheap energy and handy resources. The models so painstakingly developed over three hundred years to create more jobs and more goods and services must be dramatically modified.

What to do? First of all, we must better practice humility, better appreciate what we don't know (ignorance), and develop a culture of limits. We in the American West can help give birth to the new world that will be required on the downslide of the Hubbert curve. The West has had its own clash of cultures that tracks the knowledge/ignorance dilemma: that between the culture of the infinite and the culture of the finite.

Civilization has triumphed in the West because it has refused to accept limits and has overcome myriad obstacles. Our ancestors found a desert and made it into a garden. The culture of the infinite teaches that knowledge, ingenuity, and imagination can prevail over any obstacles and that there are no limits—only a lack of creativity.

This is the West of irrigation canals, transmountain diversions, pivot sprinklers, and other adaptations that allow us not only to live in a semidesert but also to enjoy green lawns and prosperity. The culture of the infinite suggests that the future is a logical extension of the past, that all problems have achievable solutions: "Go forth and multiply and subdue the earth" and "Go West, young man."

It is the optimism of "Not to worry: God gave man two hands and only one stomach." It reflects a devout belief in limitless economic development, progress, and the perfectibility of the human condition. It is the world of the green revolution that has given us the potential to eliminate hunger and of technology that some say has repealed the law of supply and demand and discovered endless and unlimited wealth. This is the world built around unlimited people—unsatiated consumers.

The supporters of the infinite are either the modern prophets or the modern alchemists—but, to date, they say that they have been stunningly successful in solving the problem of population and poverty. And, in their minds, their approach will continue to be successful. Knowledge trumps all! Aridity can be solved by desalinating oceans, and wealth (computer chips) can be created out of sand.

The second culture is the culture of the finite. The American West also teaches that we must adapt to nature and be acutely aware of its fickleness and limitations. It teaches us humility and caution, that there is such a thing as carrying capacity, and that we must respect the fragility of the land and the environment. It argues that nature teaches us that we never can or should rely on knowledge or the status quo, that climate is harsh and variable, and that the price of survival is to humbly anticipate and prepare and to recognize that we can never anticipate all the surprises that nature has in store for us. It questions the proposition that growth, population or economic, can go on forever. This is the world of conservation, national parks, wilderness legislation, crop rotation, Planned Parenthood, Malthus, and the *Exxon Valdez*. It is the vision of Thomas Berry: "The earth and the human community are bound in a single journey." And it listens to Isaiah: "Woe unto them that lay field upon field and house upon house that there be no place to be left alone in the world" (see Berry 1990, 150–58).

Our industrial civilization is built on the assumptions that there are no limits, that technology will forever solve our problems, and that we will not reach any sort of carrying capacity. It assumes infinite resources, that scarcity is caused by want of knowledge and imagination. Civilization in most of the world supports this assumption of the infinite.

The finite culture, with fewer adherents, but equally passionate, contends that the first culture is making empty-earth assumptions that cannot be sustained. They point out that, the more we know, the more we don't know and that technological hubris is dangerous to our future.

They want to move now to stabilize U.S. population, reduce our energy use, and help the rest of the world do likewise.

Ultimately, finite-culture adherents feel that we cannot and should not have an America of 500 million living our consumptive lifestyles. They contend that we live in a hinge of history where society must rewrite the entire script. If they are correct, then our basic assumptions about life, our great religious traditions, and our economy are conceptually obsolete. So far, those who sing this song are failed prophets.

But what if—just what if—the culture of the infinite was only a temporary victor? What if nature bats last? What if the real lesson we should have learned in a place with thirteen inches of rain was the need to appreciate that limits could be pushed and extended but never eliminated? What if the rain forests, the dying coral, and the rising temperatures are trying to tell us something?

The lessons I have learned from my fifty-year love affair with the West support this second culture. I came to Colorado stationed in the army in 1957 and immediately fell in love with its deserts, its mountains, and its stark beauty. The West accepts people for who they are, not where they were born. The West with its desert landscape transformed me; in its infinite beauty I found finiteness. I believe that we need to transform society from an earth-consuming technological civilization to a sustainable and more benign civilization. I'm impressed with Aldo Leopold's "land ethic," which teaches that human fate depends on our ability to change the basic values, beliefs, and aspirations of the total society (see Leopold 1986). I believe that the fate of the world depends on our ability to know when to abandon the infinite culture and shift to the finite culture. Wait too long, and we are doomed. Some will say that, if we shift too soon, we'll give up a lot of fun and exhilaration. I'd rather we shift too soon. We won't get a chance to shift too late. Let me end by quoting Howard Nemerov, a former poet laureate:

> Praise without end for the go-ahead zeal
> Of whoever it was invented the wheel;
> But never a word for the poor soul's sake
> that thought ahead, and invented the brake. (Nemerov 1991)

Our society needs some brakes because we are not nearly as smart as we think we are. All vainglory to the contrary, we are inextricably children, not of history, but of nature.

REFERENCES

Bartlett, Al. n.d. "Arithmetic, Population and Energy: Sustainability 101." University of Colorado at Boulder. Videotape. Available through www. cubookstore.com.

Berry, Thomas. 1990. *The Dream of the Earth.* Reprint, San Francisco: Sierra Club Books.

Boulding, Kenneth E. 1974. *Collected Papers of Kenneth Boulding.* Boulder: University Press of Colorado.

Leopold, Aldo. 1986. *Sand County Almanac (Outdoor Essays and Reflections).* Reissue ed. New York: Ballantine.

———. n.d. *Sand County Almanac Fact Sheet.* Available at www.aldoleopold. org/About/almanac.htm.

Lovins, Amory B. 2004. *Winning the Oil Endgame.* Snowmass, CO: Rocky Mountain Institute.

Morrison, Reg. 1999. *The Spirit in the Gene: Humanity's Proud Illusion and the Laws of Nature.* Ithaca, NY: Cornell University Press.

Nemerov, Howard. 1991. "To the Congress of the United States, Entering Its Third Century." *Trying Conclusions: New and Selected Poems.* Chicago: University of Chicago Press.

Royal Society and the National Academy of Sciences. 1997. Joint Statement on Population Growth, Resource Consumption and a Sustainable World. Available at www.nasonline.org.

Santayana, George. 1905. *Life of Reason, Reason in Common Sense.* New York: Scribner's.

United Nations Conference on Environment and Development, Rio de Janeiro. 1992. Rio Declaration on Environment and Development. Available at http://www.unep.org.

Youngquist, Walter. 1997. *Geodestinies: The Inevitable Control of Earth Resources over Nations and Individuals.* Portland, OR: National Book Co.

IGNORANCE AND KNOW-HOW

Conn Nugent

FUELS AND PROPHECY

In order to demonstrate ignorance, of course, we have to proceed from what we think we know. After thirty years of ingesting books, papers, and presentations, plus general observation, I think I know some things about carbon fuels and their consequences.

I know that modern societies rely—utterly, transformatively—on carbon fuels. I also believe that the consumption of those fuels will continue to increase for decades to come, that their prices and costs will increase more or less steadily, and that those prices and costs will eventually preclude the operation of economies as vast or as productive as the ones we have today. I believe that people do not know, and will not know within this century, how to maintain multibillion human populations with the purchasing power to which we are now accustomed. We are too ignorant.

So I offer some flat predictions. These are not bold predictions, but they are trustworthy. In fact, I would wager large sums on them. To each of these predictions is appended a statement that is, I believe, more likely true than not. I would entertain side bets on these.

Sure Thing. Global demand for petroleum will increase until it costs at least twice as much as it does today.

Good Bet. Petroleum will cost perhaps three or four times as much, adjusted for inflation. We will not develop alternative fuels nearly so efficient or convenient as gasoline or diesel, nor will we develop attractive alternatives to petroleum as a feedstock for plastics. Increased demand from Asia will outweigh conservation in Europe and North America.

Sure Thing. Global demand for oil will dependably exceed supply—year in, year out—no later than 2025. By the end of the century, more than 90 percent of all recoverable oil deposits will have been consumed. The price of petroleum will rise so high that gasoline, diesel, and other oil fuels will be sold as elite niche products, not so different from what they were in 1900.

Good Bet. The time required for the transition from cheap petroleum to expensive petroleum will take longer than predicted in some recent extrapolations about peak oil. As prices climb, they spur the discovery of new reserves (mostly deepwater), render economical oilshale deposits hitherto regarded as not worth the bother (Alberta and the Orinoco Basin), and spread new extraction technologies that rejuvenate wells thought to have been tapped out (Russia and Mexico). But petroleum *is* finite. Oil supplies may not have reached their peak yet, but you can certainly see the peak from here. When will oil become irrevocably expensive, that is, four times as costly as today? I'll take even money on 2040–2060.

Sure Thing. Natural gas will follow the same general trajectory as oil, but at a markedly slower pace.

Good Bet. That pace will be slower by about thirty years. Since agriculture is the most gas-intensive sector of the economy, and since governments are likely to subsidize food production, there will be strong upward price pressures for gas used in other sectors. Gas burned to generate electricity will stay in high demand in those nations committed to antipollution standards and the reduction of carbon dioxide emissions.

Sure Thing. The global demand for coal will increase for the next hundred years.

Good Bet. As it was in the period 1800–1920, coal will again be the world's dominant fuel, largely because of growing demand for electricity in developing countries.

Sure Thing. Much more coal will be liquefied and gasified.

Good Bet. As oil and gas become expensive, coal will be transmuted to do their work. The substitution process will not happen quickly, however, given the inefficiencies and higher emission levels of liquefied and gasified coal compared to oil and gas. Coal and coal-based liquids and gases will themselves become expensive if toxic emissions are scrubbed and carbon dioxide is "sequestered" so as to mitigate the impact of coal burning on atmospheric warming. Keep coal cheap, and you speed the ruination of the natural world. Make it behave politely, and it gets expensive.

Sure Thing. Nonfossil sources of energy—primarily nuclear power and solar power—will increase their shares of global energy supply, but none will take over completely.

Good Bet. Many rich countries will follow the French model of using nuclear power as the primary source for generating electricity. Wind turbines and photovoltaic arrays will make significant contributions in countries uneasy with nuclear power but dedicated to lowering carbon dioxide emissions. Biofuels will play a modest role in the tropics.

Sure Thing. The rate of increase of aggregate temperature on the earth will accelerate over the next twenty-five years. The overall warming trend will continue for another sixty years at least, during which time the average annual temperature of the planet will increase by at least four degrees Fahrenheit and sea levels will rise at least three feet. An increasing share of private and public expenditures will be spent coping with the effects of climate change on human activities.

Good Bet. Those effects will vary widely. Growing seasons will stretch in Russia and northern North America, and plant growth will be enhanced in many places. On the whole, however, negative effects will much outweigh the positive. "Dramatic weather events" will increase. Prolonged droughts will afflict farmers in the lower latitudes, especially in Africa. The flooding of coastal areas will increase, and low-lying settlements too poor to build seawalls will be devastated (Micronesia, most of Bangladesh). Tropical diseases will spread to once-cooler regions. Agricultural expansion and environmental disruption will expose humans and farm animals to new pathogens.

Sure Thing. Global food production will increase over the next twenty-five years, thanks to the spread of high-input agriculture to more regions of the developing world. Soil loss and soil degradation will increase as a result. Hypoxia from fertilizer runoff will form extensive "dead zones" in the waters off the coasts of every continent. Aquifers will be depleted and water supplies salinated. Claims to irrigation rights will foment domestic and international disputes.

Good Bet. For a while, at least, green revolution–style innovations will help agriculture keep pace with a human population growth now expected to peak at 9.1 billion in 2050 (the middle-range projection of the UN Population Fund). Government subsidies for food production will postpone the effects of rising natural gas prices on the use of synthetic fertilizers. The greatest threats to food production will come from the ecological degradation intrinsic to modern farming methods, especially as they are applied to soils that are thin to begin with. Benign alternatives to high-input farming—no-till, perennial cereal crops, for example—will not emerge commercially until 2020 at the earliest.

THIS GOLDEN AGE

All terrestrial species need fresh infusions of carbon in order to survive. We are always on a carbon hunt. For most of human history, our species acquired its carbon by eating plants and animals on a more or less replenishable scale. About ten thousand years ago, humans devised ways to organize the carbon seeking through the domestication of plants and animals. They were able to form larger, less migratory populations with the capacity to generate surpluses sufficient for the emergence of classes that did not have to produce food. These capacities came at the cost of chronic soil loss and other forms of natural resource degradation. As Wes Jackson says, "That's not a problem *in* agriculture. That's the problem *of* agriculture."

Over the last two hundred years, humans have exponentially multiplied their carbon acquisitions by learning to extract the energy contained in the buried remains of hundreds of millions of generations of plants and animals (hence "fossil" fuels). This new capacity enabled enormous growth in production and the share of the population uninvolved with providing food. It also occasioned a consequent growth in environmental degradation. That is the problem of industrialization and, later, industrialized agriculture.

The wide adoption of coal burning marked the beginning of the fossil fuel era. But perhaps the most socially transformative development was the first commercially successful petroleum well, drilled in Pennsylvania in 1859 (the same year as the publication of *On the Origin of Species*). It took about fifty years for King Coal to give way to King Oil, but petroleum has reigned ever since. Its popularity is growing still and at an astonishing rate: one-quarter of all the oil ever burned has been burned since 1994; half of all the oil ever burned has been burned since 1980.

Among our oil dependencies, the strongest attachments are to portable petroleum fuels. And the reason we are so attached is because they are truly wonderful.

Consider gasoline, petroleum's most refined offspring. Gasoline is safe, even when dispensed by amateurs at the self-service station. It is efficiently concentrated: an ounce of gasoline contains more potential energy than an ounce of any other fossil fuel and, consequently, makes available more vehicle space for passengers and cargo. Gasoline is easily stored and transported and doesn't have to be recharged, reconstituted, or re-anythinged. After combustion, the most noxious of its emissions are rendered relatively harmless by an inexpensive catalytic converter. It is by far the most convenient of fuels.

Gasoline (and diesel) enables mass mobility, and the lure of mass mobility has proved irresistible in every society that could afford it. So long as we prize mobility—for our goods as well as for ourselves—we will buy gasoline until its price becomes just too onerous to bear. At some point, high gasoline prices might move us to other forms of vehicle power, and we might even choose to move around less. But price signals will have to be bright and unambiguous. Gasoline is just too good.

Natural gas is not quite in gasoline's league as a portable fuel, but it is certainly versatile and altogether handy. Natural gas is dense and burns relatively cleanly. It can be transported as a gas through pipelines and as a liquid in tanks. It can power vehicles and electric generators; it can heat buildings; and, most important, it provides the feedstock for the synthetic fertilizers on which industrial agriculture relies.

Modern agriculture, in fact, is enmeshed in a web of cheap oil and natural gas. Farmers need off-farm, imported carbons for the nitrogen fertilizers that compensate for their annual losses of soil and fertility; off-farm carbons to power the machines that plant, cultivate, and harvest; off-farm carbons for herbicides and pesticides; and off-farm carbons to transport products to consumers. This system works as a successful strategy for increasing food supplies. But it works only until the prices of off-farm carbon go high or until soil depletion precludes any kind of agriculture at all.

It is hard to find *any* sector that would be unaffected by a meaningful rise in the prices of oil and gas. Closely examine a slice of a modern economy, and you recognize that the physical and human infrastructures—the buildings, the machines, the supplies, the transport, the life supports of workers—are created by burning carbon fuels. Make carbon fuels expensive, and those infrastructures become expensive, both to build and to operate. This is true not just for farms and factories but also for offices, schools, and homes.

There have been some interesting recent attempts to quantify the materials and energy embedded in everyday objects and behaviors. The best known is probably the carbon footprint, a calculation devised by Canadian scientists to measure carbon dioxide emissions occasioned by typical North American consumers. Basically, you estimate how much carbon went into the making of a product, the transportation of the product, the use of the product, and the disposal of the product. Your breakfast cereal requires x amount of carbon for growing the grains (fertilizers, tractors, combines, fuels), y amount of carbon for processing (trucks, truck fuel, mill operations, packaging), and z amount of carbon for moving it to your supermarket, and then your kitchen table, and then the landfill. The footprints of certain activities, like eating and commuting, can vary according to personal custom, as in where you shop and whether you ride a bus to work. The footprints of other kinds of activities, like jet travel, are pretty much set: once you decide to fly somewhere, there's not a lot you can do to modify your demand.

Yet research shows that households that make a small carbon footprint in one realm tend to make a big footprint in another. At least, in this time, and on this continent, the size of a household's carbon footprint is directly correlated to its overall expenditures. Spend money, and you burn carbon. Spend lots of money, and you burn lots of carbon. A service economy of educated symbol manipulators is still intensively material.

But it's been a great ride, has it not? Cheap carbon has allowed a billion people to lead lives available to only the thinnest upper crust of two hundred years ago. Most contemporary North Americans, West Europeans, and Japanese obtain food, shelter, transport, material goods, and medical care through labor that requires about one-quarter of one's time and little physical pain. The suburban grandchildren of peasants and mill workers enjoy the luxuries of mini-Versailleses—and with better plumbing. This state of affairs has held steady for those of us born into the postwar American middle class, and it is available today to ambitious and lucky Asian teenagers. Jump into global capitalism, and the chances are good—maybe not great, but good—that you and your family will live longer, eat better, and acquire more possessions than your parents did. The Asian teenagers know that petroleum is finite and due to become more expensive. But they would rather face that eventuality as rich people than as poor people. As a Chinese energy official told an American engineer recently: "As soon as anyone develops a clean way to generate power, we'll be the second ones to use it." He is not wrong, not yet. We are still in the golden age.

THE CHILDREN

So we have some idea of the nature of an upcoming sea change in the material circumstances of our descendants. Will they manage? Will they *know* enough to manage?

Americans set great store by knowledge, and we assign special value to knowledge that expands productive capacities. The exercise of those capacities—knowledge turned to work—we call *know-how*. We praise knowledge, but we treasure know-how. In fact, knowledge unattached to know-how ("mere knowledge") we tend to regard with amused tolerance or open suspicion. We also think we know what kinds of knowledge lead most efficiently to the kinds of know-how we regard most highly. We tend to promote and reward those kinds of knowledge, lately under the banner of "preparing our kids to compete in a global economy."

But we do so in ignorance, I believe. We are ignorant—innocently and willfully—of the degrees to which the character and utility of our know-how are predicated on the availability of cheap carbon. The value of certain kinds of know-how that we now deem essential will decline as the costs of carbon rise. Other kinds of know-how will become more valuable.

It's not simply a matter of rising demand for wind turbines and declining demand for V-8 engines. The general sets of know-how required to prosper in an international system based on cheap access to distant energy sources are different from the sets of know-how you need to prosper in a local system where energy is expensive. A large-scale spinach grower knows things of which an Amish farmer is ignorant, and vice versa. Today's common cultural know-how assumes conditions familiar to the large spinach grower, but those conditions are now revealed to be impermanent. The Amish farmer suddenly gets smarter.

He has know-how useful to a society that needs to feed itself without immense amounts of purchased carbon. I don't doubt that our grandchildren will get most of their food from farms with computers and tractors—the most valuable kind of know-how will probably be a blend of traditional knowledge and ultramodern knowledge—but the giant high-import farms of today should become increasingly unprofitable, and the types of know-how they summon should become decreasingly relevant.

As oil and gas become expensive, the production of goods that are transport intensive will be disadvantaged, as will production that relies on high carbon-energy inputs for the creation of the goods themselves. Enterprises of far-flung suppliers and far-flung markets will need to adjust. Local low-input producers will appeal to new markets. As that trend grows, so will the probability that, relative to today, a larger share of the population of rich countries will produce goods and a smaller share will sell services. If so, that would constitute an important reversal of the heretofore irresistible momentum of developed countries toward the service economy model first described by Daniel Bell forty years ago. I worry that we are too ignorant to handle the change.

First of all, I assume that most consumer goods will cost more than they do today, as measured by per capita purchasing power, if for no other reason than because of a general rise in material and energy costs. I would also assume that our grandchildren's societies will probably suffer from at least an initial scarcity of cultural capacity to cope with a decline in purchasing power (however slight) and to reinvigorate and manage local and regional economies. A fundamental difficulty is that cheap-energy know-how has atrophied expensive-energy know-how. Just as pre-Gutenberg savants had powers of memory and recitation not found in the print era, pre-oil-and-gas farmers mastered agronomic and mechanical systems alien to their descendants but of potential utility in developing the optimal 2030 toolbox that hybridizes old and new approaches to farming. New circumstances will favor skill sets and attitudes that once flourished in harder times—knowing how to fix things, for instance—but have latterly fallen into disuse.

OK, we moderns say, so we've lost some old-timey know-how—we haven't *lost* it, actually, it's more like we've *mothballed* it—but, all in all, it's a fair exchange. We know things our ancestors couldn't imagine.

But perhaps we underestimate the value of old know-how and the difficulty of a rediffusion of it among the populace. There are scores of historical examples where high-level skills once common became rare or debased. J. B. Bury, the great scholar of the ancient Mediterranean, observed that it took only three generations for a people to lose a craft. And maybe we overestimate the applicability of current know-how. Most of us are aware that mobility depends on carbon fuels, and most of us are aware that carbon fuels power industry, but few conclude that our knowledge is *embedded* in cheap carbon fuels and may be of scant use in another energy context. It could be that the person best prepared for the age to come is the skeevy nineteen-year-old who drives a pickup filled with wires and machine parts that he's going to try to rig up next week. Today he's a disappointing dropout, ill equipped to compete in that global economy we talk about. Later on he's a very useful guy to have around.

THE MOST IMPORTANT QUESTION ON EARTH

When I discuss these fuel and food questions with nonspecialist colleagues, their reactions often tend to a sort of humanistic uplift. Don't worry so much. Don't underestimate ingenuity and the human spirit. Science and technology will respond in ways yet inconceivable to our blinkered imaginations. Don't tell me you're one of those neo-Malthusians.

Yes, human ingenuity is powerful and surprising. Few seers of 1900 foresaw airplanes or antibiotics. But there's a huge difference between new technologies and new energy sources. We can be dazzled, even changed, by new machines and new processes, but the form of energy on which they depend is usually familiar. New, transformative sources of energy don't appear very frequently. Hans Bethe used to point out that no form of energy—from the draft horse, to coal, to petroleum, to nuclear power—ever became a fuel for commonplace technology in fewer than fifty years. Bethe said that the first half of the twenty-first century would be powered by energy sources familiar to the scientists of the second half of the twentieth century. Cold fusion, anyone?

I fear that the post-2030 world will be extremely difficult for hundreds of millions of people. Climate change, plus ecosystem destruction, plus soil depletion, plus high fuel and fertilizer prices, plus continued population growth—it reads like a recipe for famine and disease. And it would be surprising if competition for depleted resources did not contribute to, or impel, wars and banditry and social upheavals of all sorts.

That is not to say that everyone everywhere will feel imperiled, much less uncomfortable. Rich people and rich countries will protect themselves. Money will be more useful than ever. I expect that the manner of living enjoyed today by the middle classes of the developed world will hang in there for a few more decades at least and probably longer. If our national governments can resist the temptation to use military force to secure energy supplies, if they and we can manage our fiscal and economic affairs to avoid a sudden plunge into a new Great Depression, and if the weather doesn't turn completely haywire—big ifs—life for most of us Americans in the twenty-first century might be as rewarding and as miserable as it is today. It's not a material issue per se: people have led healthy, complicated lives on one-tenth of our current per capita energy diet, and people can do it again. It's more a question of our social and personal reactions to the new expensive-energy environment. Will an increase in fossil fuel prices reduce the size of the service economy, curtail air travel, and limit consumer choice? Yes. Will new houses be as big and as cheap to heat and air-condition as the new houses of today? No. Will our beach house wash away? Probably. Will we have our laptops and iPods? Yes, and better than ever. Will more people be thrown back (as they say) on their own capacities—and the capacities of their neighbors—to grow, make, and repair? Yes. It might not be so bad.

It might even be good. Already a significant minority has found that the pursuit of happiness is well taken on a path marked by energy conservation, local food, and a general attentiveness to the natural world. I am referring not just to the green counterculture—though it certainly counts—but to all sorts of people: new traditionalists attracted to rural life and agrarian values; cyberphiles pioneering the hybridization of old and new know-hows; settlers of energy-efficient urban neighborhoods; Mexican workers balancing on one foot in Chicago and one foot in Oaxaca. Just as it's conceivable that the giant flying apparatus of the global economy will crash and burn, it's also conceivable that Americans will learn how to adapt, even flourish.

I can imagine a federal government that hastens the development of renewable energy as the great public works project of our time. Solar trains speed you from New York to Obama City (formerly San Francisco) in under fifteen hours. All new buildings incorporate photovoltaic arrays. The Dakotas produce a bumper crop of Lutheran "wind-millionaires." I can also imagine a fruitful repopulating of the North American countryside. I can imagine a new commonwealth of vital local institutions, synthesizing the voluntarism limned by de Tocqueville and the ideal of universal human rights that has animated our own times.

I am more worried about poor people in poor countries. No soft landing for them, no character-building belt-tightening. My basic fear is that a tropical agriculture strategy successful in the first third of the century will—because of high fertilizer prices, soil loss, and the baneful effects of climate change—undergo a terrible decline in productivity just as the number of human souls exceeds 9 billion.

The most fundamental need of the poor, and of all of us, is the same: food. What poor farmers need in a matter of decades—and what everyone else will need sooner or later—is a high-yield agriculture that does not rely on off-farm sources of energy and fertility. One would suppose that major research institutions are doing serious work on this. But they are not. There is no large-scale program anywhere to inquire into farm regimes that could run themselves exclusively on contemporary sunlight. Only a handful of scientists—botanists, plant geneticists, plant breeders—are pursuing the most pressing question on earth: Can we develop food systems that produce high yields, provide their own fertility, build their soils, and handle pests and diseases with the resilience of a natural ecosystem?

And that most pressing question might be regarded as the introductory assignment for the most important question: Can we devise support systems that provide us with the means to gain sustenance and shelter, promote health and longevity, and afford ample opportunities for free expression without depleting the common stock of natural resources? Or as Angus Wright once put it: Can we make a world where conservation is a consequence of production?

Part Two

Second Cut

OPTIMIZING UNCERTAINTY

Raymond H. Dean

We instinctively modulate our boundaries, expanding them to expose new options and contracting them to cull out poorer options. As boundaries expand, more accessible information raises the level of complexity, and we must pay more careful attention to understand that higher complexity. As boundaries expand beyond our ability to comprehend, danger increases.

Wisdom tells us to set our boundaries so that we can just comprehend the accessible information within those boundaries. Weak or distant boundaries expose us to more information than we can digest. Poorly digested information is perverse ignorance. As boundaries expand, that perverse ignorance grows into a strange distortion of reality—a distortion that can seduce us into serious error.[1]

Nature and culture automatically react to inadequate boundaries by building new ones. Unless we restrain ourselves in our manipulation of the natural world, that world and the social, political, and economic world within it will impose restraints on us.

GRASPING THE SITUATION

Scientific development changes the way people think about the human situation. Copernicus's description of planetary motion removed humans from the center of the universe. A century later, Isaac Newton provided proof that the knowable universe was orderly. But two centuries later, Darwin's description of biological variation and the inherent uncertainty of quantum mechanics muddied the waters again. Then we began to think of the orderly patterns we see as emerging from randomness, and this way of thinking permeated biology, ecology, the social sciences, and communications.

In the course of this development, we learned that a collection of data and a collection of energetic particles share a common property called entropy. This common concept of entropy provides a basis for an analogy among different scientific and cultural domains.[2] Also, Shannon, Nyquist, and others independently discovered an important information-processing principle called the sampling theorem. To faithfully reconstruct original information, the receiver of incoming data must sample at a rate no less than twice the bandwidth of the analog filter through which that data flow. Lower sampling rates distort the original information in a strange way—like the images created by a strobe light at a rock concert or the occasional impression that wheels on a car are rotating the wrong way.[3] In other words, if our boundary of consideration (sampling rate) is smaller than our physical boundary (filter), our understanding is not just incomplete or obscured by random noise. It's deceptively perverted into strange patterns.

Toward the end of the twentieth century, we begin to notice that many other supposedly random phenomena are not actually random. Agents are interconnected. Older agents accumulate more interconnections, and now we see optimized interconnection networks[4] in nested patterns that look like fractals[5]—patterns that crowd into relatively small portions of the available space and repeat at different levels of scale.

Thus, we had a deterministic thesis, followed by a random antithesis, followed by a synthesis. In the synthesis, *we see that the inorganic world and the organic creatures within it are interconnected in complex ways and dwell in an uncertain state between order and disorder.* They naturally seek a balance between no alternatives (too little information), on the one hand, and excessive confusion (too much information), on the other.[6] The balance provides just the right awareness of plausible opportunities and threats—optimized available information.

However, many would say that, instead of trying to optimize available information, it's better to try to maximize it. They imagine that maximizing available information also maximizes control. But sometimes we get carried away. We invent gunpowder, dynamite, and atomic weapons. We extinguish species. We exhaust natural resources that took millions of years to accumulate. We wash away the soil that grows our food, poison the water we drink, and pollute the air we breathe.

In spite of these pernicious effects, our faith in technology—"modern optimism"—persists. We believe that we'll get the best of all possible worlds if we just do more of what we have been doing—gather more information, use faster communication to disseminate it, and depend on free enterprise to produce the best results. Fast communication and cheap oil have made it easy to employ cheap labor in remote parts of the world, and governments have removed trade barriers. Thomas Friedman says that now the world is "flat."[7] *Flat* doesn't mean physically flat. It means a level playing field, a shallow hierarchy—everyone has access to everyone else's territory. This increases local options, but it also decreases local stability.

Flat-world apologists claim that the loss of local stability is morally justified because it increases average wealth. It "raises all boats." They justify their faith in education, communications, and economics by citing the historical correlation between growth in scientific knowledge and growth in global wealth. There may be a relation between knowledge and wealth, but many flat-world enthusiasts seem to ignore the support that both get from the rapid consumption of ancient fossil and nuclear fuels, biological diversity, and good soil. When these resources run out, instead of raising all boats, the turbulence of flat-world economics might sink all boats.

Instead of using Friedman's simple "flat" metaphor to explain the current state of the world, I think it's better to think of the world as a collection of interconnected leaky balloons, with one balloon for humans and their artifacts, another for other animals, another for plants, another for microbes, another for fossil and nuclear fuels, etc. The sun pumps up the plant balloon. For the rest of us, plants are the "primary producers" because they transform the sun's electromagnetic energy into forms of chemical energy that other creatures can assimilate. Animals get their energy by eating plants or other animals. Microbes and geologic forces transform some of the energy in previously living plants and animals into coal, oil, and natural gas. Humans get their energy from these other system components.

To increase flow into the human balloon, humans enlarge the connections between the human balloon and the other balloons and then punch holes in the human balloon to increase its leakiness. Increasing leakage through the holes in the human balloon reduces its pressure, and this induces more flow from the other balloons into the human balloon. Because it supports this enlarging of connections and punching of holes, the unbridled pursuit of knowledge sucks energy from the rest of the world.

When flat-world enthusiasts say that flattening the world increases average wealth, they assume that human economics is a sufficient measure of wealth. But, when it comes to natural resources, human economics knows only the cost of extraction. It is ignorant of the inherent value in the resources themselves. When we recognize this ignorance, instead of seeing humans as wealth creators, we will see humans as the heaviest consumers of limited common resources. Then we will see that knowledge and wealth both depend on energy consumption, and we will see that our current effervescence will fizzle when the world's other balloons deflate.

Flat-world enthusiasts encourage us to go with the flow. But suppose that a future fizzle doesn't appeal to you. What can you do? You can do what living creatures have done for the last 2 billion years, ever since evolution invented eukaryotic cells—the cells in all higher forms of life. Eukaryotic cells contain within themselves an encapsulated nucleus and other encapsulated organelles. The encapsulating membranes improved biological fitness by restraining the flow of energy, material, and information.[8] Similarly, you can improve your fitness by encapsulating yourself in a balloon of restraint that restricts the flow of energy, material, and information through the part of the world you control. This will help you live more gracefully by making your world more stable.

LIMITING CONFUSION

Physical and intellectual boundaries improve stability, and they make it worthwhile for people to invest in tools, skills, and relationships, just as they make it worthwhile for plants to invest in roots, leaves, seeds, and symbiotic partnerships. You want the size and strength of your physical and intellectual boundaries to be just right. If your physical boundaries are closer and tighter than your potential boundaries of consideration, you miss opportunities you could develop. If your physical boundaries are farther and looser than your maximum possible boundaries of consideration, you misjudge things and make serious mistakes. If both these boundaries are too close and too tight, you don't have the variety of resources you need to survive and deal with unforeseen happenings. If both these boundaries are too far away and too loose, you're buffeted around and perpetually at your wit's end. Imagine a lioness looking for a meal. She comes upon a large herd of gazelle. Does she try to bring down the whole herd? No. She cuts out a few, identifies one, and then chases that one only.

The most appropriate place for your boundary of consideration depends on the situation. Suppose you're crossing the Atlantic Ocean alone in a thirty-three-foot sailboat. It's 10:00 A.M., the waves are four feet high and thirty feet long, and they're coming from the same direction as a steady ten-knot wind. The sky is clear. You're wondering what to do in the next couple of hours. Your problem is easy (too easy!) because your boundary of consideration is relatively close and it doesn't include any worries. You decide to accept your ignorance of the world over the horizon and sit back and enjoy the easy life.

Six hours later, clouds appear, and the wind shifts. New two-hundred-foot-long swells tell you that there's a gale somewhere over the horizon and it's coming toward you. Ahead lie twelve hours of night travel through heavy shipping lanes. Your problem is suddenly more difficult than before because your boundary of consideration is farther away in both space and time and it includes several things to worry about. Now, finally, you decide to reduce your ignorance by getting a weather report.

At midnight, the gale arrives, floating debris punches a hole in your sailboat, and it begins to sink. Now you must gather freshwater, food, and other supplies for 120 days and fifteen hundred miles of drifting in an open rubber dinghy, and you must launch that dinghy quickly. Your original protective physical boundary (your sailboat) is disappearing. Your problem is now very difficult (too difficult!) because your boundary of consideration (a distant landfall) is very far away and there are too many things to think about. To control your panic, you decide to ignore most of those worries and focus on immediate tasks—loading and launching the rubber dinghy. In other words, you intentionally shrink your boundary of consideration—accept more ignorance—to make your confusion tolerable.

Boundaries play a major role in evolution. Consider Darwin's finches—a group of birds on the Galapagos Islands that Charles Darwin studied. When these birds' ancestors migrated to these islands from South America, the physical distance between the islands and the continent separated them from their parent population.[9] This open-water boundary made these birds ignorant of potential mates on the distant mainland. Reproductive isolation permitted the genetic characteristics of the relatively small pioneering populations on the islands to drift in time as mutation and selection adapted them to local conditions. Eventually, such drifting forms a new species, and this enriches the world's biological diversity.

To maximize their individual fitness, individual humans—like finches—naturally focus on whatever knowledge is most relevant to them personally. It's not the body of all human knowledge that matters. It's the individual bundles of knowledge possessed by small groups of human beings. These individual bundles of knowledge are limited to the capacity of a few human brains. The genetics that affected the evolution of Galapagos Island finches was the genetics carried by just a few birds. Similarly, the human knowledge that determines our children's futures is knowledge that can fit into the heads of just a few of us. Let's start by considering what one individual can perceive.

PERCEIVING ONE'S SURROUNDINGS

Look at figure 1. The heavy dotted line represents you. The white area represents your system—the things, creatures, and information you can access. It takes energy to produce and sustain many of the things and creatures in your system (including you). With unavoidable boundary leakage, the information in your system changes in time, so you must keep sampling to stay up-to-date. This sampling activity also consumes energy. So you need a flow of energy to sustain everything in your system, including you and the information you can access. This energy flow maintains a special kind of energetic capability, which classical mechanics calls a virial,[10] so you can think of the white area alternatively as representing your system's energetic capability.

The light gray labels identify boundaries that separate things and information you can access from things and information you cannot access. If your information-sampling rate is sufficiently high relative to the size of the bounded area, when you look at the boundaries, you see plain blank walls, and you think, "I don't know." That's benign ignorance. If your information-sampling rate is too low relative to the size of the bounded area, when you look at the boundaries, you see them painted with strange images,[11] and these images distort your thinking. That's perverse ignorance. Dark gray areas in figure 1 represent inaccessible things and information. The gray blocked arrows represent forces you do not perceive.

For each oval in figure 1, envision a third dimension, perpendicular to the plane of the page. This third dimension is relative time.[12] The plane of the page is the present, what's above it is the future, and what's below it is the past. The hemisphere above the page is filled with continuous probability estimates of future possibilities. The hemisphere below the

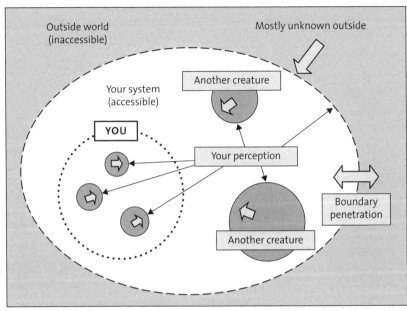

Figure 1. Graphic representation of the range of individual perception. The plane of the page is the present; white space above the page is the anticipated future; white space below is the remembered past. Typical real boundaries are not smooth curves like those shown. They are ragged fractals, but those fractals still enclose bounded areas.

page is filled with discrete remembered historical events. Think of each oval in this picture as a slice through the middle of a constant-volume balloon.[13] You can enlarge the distances in the plane of the paper (flatten your world) by squashing the balloon, but this simultaneously shortens the distance perpendicular to the paper—your time horizon. Thus, flattening your world shortens the time available for you to derive benefit from previous investments in tools and relationships, and it forces you into less efficient behavior.[14]

What you perceive includes some of the world outside you, and it includes some of your inner self, although it's easier to see outward than inward because larger-scale boundaries (dashed lines) are typically more porous than smaller-scale boundaries (solid lines). You are aware of other members of your community, and you can interact with them. You are aware of parts of yourself, and you can feel and/or control those parts. On the other hand, you are ignorant of much of what's inside those parts, like your subconscious. You are also ignorant of what's inside other members of your community. You are also mostly ignorant of what's outside your community. Nevertheless, all real boundaries have some openings, and the double-ended light gray arrow depicts perceived flow through known openings in your system's boundary.

Look at the boundary between your physical and intellectual system (in white) and the outside world (in gray). You have some power to adjust this boundary by expanding or contracting it to include more or fewer other creatures, inanimate entities, and ideas. You also have some power to change the permeability and strength of this boundary and, thereby, alter communication with the outside world. These powers enable you to alter your system's stability and lifetime. Adequate periods of tranquility—some slack in the schedule—is an important aspect of fitness. We all need sleep, for example.

You get more slack time with a more impervious external boundary that reduces outside-world communication. You reach equilibrium more quickly with a smaller boundary that reduces internal complexity. So, if you want to know your compatriots well and perfect the quality of your tools and techniques, you want a relatively small and impervious boundary. But, if your boundary is too small and too tight, you won't have enough compatriots, tools, or techniques to meet your needs.

Therefore, you should *adjust your physical and intellectual boundaries to include just as much as you can develop and use in the time available.* On the one hand, you can seek boundaries that keep your system small and simple. In this case, your system can and must reach its full capability quickly. On the other hand, you can seek boundaries that make your system large and complex. In this case, your boundaries must be strong enough to give your system the time it needs to reach its full capability. For difficult problems, it turns out that the critical region between where there is no solution and where there are too many possibilities is extremely narrow.[15] So, for difficult problems, you must set your physical and intellectual boundaries very precisely.

WORKING WITH UNCERTAINTY

As your environment changes, the best places for your physical and intellectual boundaries also change. To solve a problem, expand or contract your thinking to find the critical region where it looks like there will be a *few* solutions—more than one but not a large number. Then, *alternate between thinking "just within the box" and "just outside the box."* Thinking outside the box typically generates too many solutions and defocuses your attention, but, after a while, you can make some selections, come back inside the box, and end up with an acceptable number of viable alternatives. When you alternately enlarge and shrink a physical and intellectual boundary, you create a cycle. Figure 2 depicts this cycling process.

The dashed lines in figure 2 represent an expanded and contracted physical and intellectual boundary. The different dashing type suggests that the boundary is more open in its expanded state. The "methods" and "applications" variables on the two axes describe complementary attributes of human problem solving.[16] "Available time" comes out of the plane of the paper. The white volume inside the surface defined by

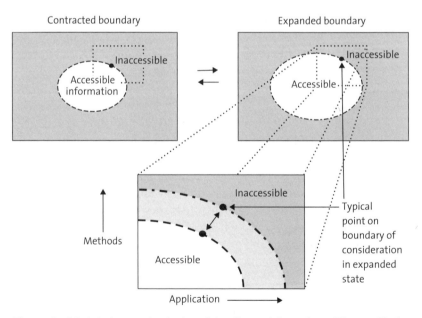

Figure 2. Modulating a physical and intellectual boundary. The oscillating dot represents a particular point on the boundary as it expands and contracts. "Methods" is the number of equally probable alternative ways to solve a particular problem; "applications" is the number of alternative problems that can be solved in a particular way; "available time" rises out of the paper. The light-gray area is your uncertainty.

the dashed lines and your time horizons is what you know. The dark volume outside is what you don't know. Moving the boundary outward and upward loosens constraints and increases alternatives by increasing the number of alternative methods, the number of applications, and the available time. This increases options, but it also increases the amount of information you must consider. Eventually, it raises your confusion and the risk of making a serious mistake to an intolerable level.

How does this picture relate to education? Is education pushing the boundary continuously outward? Pushing the boundary outward brings progressively more possibilities into consideration. But, eventually, pure data acquisition becomes counterproductive because it increases confusion. We also need to consolidate and integrate. In other words, we must eventually contract our boundary of consideration and refocus. That's when we discard poorer ideas and select better ones. This culling generates understanding and solves problems. It turns on the light and makes us say, "Ah ha! That's it!"

This kind of cycling also appears at a larger scale in human history. Here are two examples: the French and Russian revolutions exemplified alternation between tight repressive boundaries under kings (France) and czars (Russia), loose boundaries during brief chaotic interludes, and tight repressive boundaries again under Napoléon (France) and Lenin (Russia).

This kind of cycling is not just a cultural phenomenon. It's induced automatically in all living systems by the flow of energy through those systems.[17] It's also an essential aspect of evolution, where it appears as a continuous alternation between diversification and proliferation (outward boundary movement) and selection (inward boundary movement). For example, expansion occurs when fish spew out and fertilize many eggs, and contraction occurs when only a few of the fertilized eggs hatch and live to produce offspring of their own. Thus, nature and culture continuously exercise the trade-off between increasing options by diversifying (increasing "methods" in fig. 2) and proliferating (increasing "applications" in fig. 2),[18] on the one hand, and improving quality by selecting (shrinking the white area in fig. 2), on the other.

It's inevitable. When boundaries are too tight, nature and culture spontaneously loosen them. If we are too rigid, we will be disturbed. On the other hand, when boundaries are too loose, nature and culture spontaneously tighten them. If we don't constrain ourselves, we will be constrained.

COOPERATING WITH COLLEAGUES AND INTERACTING WITH STRANGERS

In spite of the value of reasonable boundaries, modern optimists want to destroy them. They want to maximize everyone's exposure and rely on global economics to produce the best result. This is a costly enterprise. It is not sustainable because it relies heavily on the rapid consumption of resources accumulated over very long periods of time—thousands of years for soil, millions of years for fossil fuels, and billions of years for nuclear fuels. Moreover, since the disturbances associated with a flattened world shorten everyone's time horizon, modern optimism drives people away from long-term cooperation and toward short-term greedy behavior, which grabs the first resource available, uses it briefly, and then throws it away. This behavior is inherently less efficient than reusing previously developed tools, techniques, and relationships. For sustainable and efficient living, we need much more physical and intellectual constraint than modern optimists are inclined to accept.

Nevertheless, few constraints are absolute, and most boundaries can be penetrated. All viable systems must have openings in their boundaries to admit nutrients (energy, material, and information) and discharge wastes. Through these and other openings in their boundaries, members of an organization and whole organizations interact with peers, subordinates, and outsiders. Now let's consider two aspects of these interactions, both of which affect the fitness of all members of any system. The first aspect is cooperation among colleagues. The second aspect is interaction with strangers. The questions to answer are: How should we view these interactions? How should we engage in them?

First, consider cooperation among colleagues. Cooperation within a local system increases the fitness of individuals in that system. Because such cooperation helped their ancestors, humans and other species have evolved dispositions and modes of behavior that support local cooperation.[19] This includes a disposition to accept group mores, a willingness to make sacrifices that benefit the group, a willingness to accept the dangers involved in punishing those within the group who do not cooperate, and a willingness to help defend the group from outside attack. *When there is effective cooperation in a local group, the group becomes a kind of superorganism and presents a coherent face to the outside world.*

Such an organization may coexist with other organizations in some larger system, as shown in figure 3. Compare figure 3 with figure 1. In figure 1, the heavy dotted line represented an individual. In figure 3, the heavy dotted line represents an organization. The basic structure in figure 3 is the same as the basic structure in figure 1. The analogies between information systems, ecological systems, and thermodynamic systems apply to figure 3 just as they apply to figure 1. Just as in figure 1, in figure 3 boundaries of consideration should extend all the way to physical boundaries to avoid the perverse ignorance that arises when they do not. The modulation effects shown in figure 2 also apply equally to figure 3. In other words, these phenomena are fractals. They appear over and over again at many levels of scale—at the microscopic level, at the level of individuals, at the level of communities, at the level of nations, and—to some extent—at the level of the whole planet.

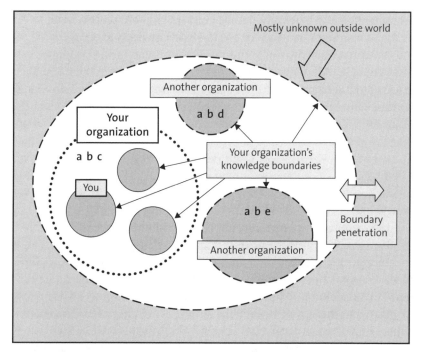

Figure 3. Graphic representation of the range of organizational perception. Just as in figure 1 above, the plane of the page is the present; white space above the page is the future; and white space below the page is the past. An added feature is indicated by the letters *a*, *b*, *c*, *d*, and *e*, which represent five different ways of doing things.

Of course, increases in scale allow new forms of these (and other) properties to emerge. For example, once we rise above the level of an individual organism, the opportunity for diversification increases. In addition to diversity in the forms of knowledge possessed by one individual, we get diversity in the individuals themselves, and, at a larger scale, we get diversity in organizations. Each boldface lowercase letter in figure 3 stands for a particular way of doing things (a "method"). In this simplistic illustration, each organization contains three ways of doing things. If all three organizations had the same three ways of doing things, the number of ways in the combination of all three organizations would be three also. But the number of ways of doing things typically increases with scale.[20] Figure 3 suggests that each organization contains one unique way of doing things (ways *c*, *d*, and *e*), so, in total, there are five different ways of doing things (ways *a*, *b*, *c*, *d*, and *e*), and the number of ways in the combination of all three organizations is five. Organizational differences facilitate specialization, and this gives the combination of three organizations an extra advantage.

Now consider the white area in figure 3—the range of organizational knowledge. This knowledge is the managerial glue that connects the a, b, and c ways of doing things within your organization and relates them to other organizations. The outer edge of the white space represents the outer boundary of the system in which your organization lives. The breadth of organizational knowledge is greater than the breadth of an individual's knowledge, but the depth of organizational knowledge is less. For example, an individual can know something about what's inside himself or herself, but an organization can know only what its individual members present to it—what it can see of them from the outside. Thus, an individual plays the role in figure 3 that one of that individual's internal organs plays in figure 1.

At any particular level of scale, organizational knowledge is not more than the sum of its parts. It is not even equal to the sum of its parts. It is substantially less than the sum of its parts. That's because an organization to which you belong cannot know everything you know. Like it or not, an individual has a private life that that individual's organization can never really appreciate. Thus, an organization's knowledge is substantially less than the sum of the knowledge of all individuals in that organization. In human culture, an organization's knowledge is the knowledge in the heads of the leaders of the organization. In an ecosystem, it's the knowledge possessed by the predators at the top of the food chain. In an individual organism, it's the connections in the nervous system or the information accumulated in the immune system. In an engineered system, it's the highest level of control—like the autopilot in an aircraft.

In the world, organizational nesting repeats through many levels of scale. The more open outer boundary lines in figure 3 indicate that, as scale increases, boundaries become more permeable. Making the world flatter increases the permeability of boundaries at all scales and tends to make everything more homogeneous. This increases the number of ways at small scales, reduces the rate at which the number of ways increases with scale, and reduces the global number of ways. Regardless of flatness, however, as scale increases, the white area (managerial knowledge) becomes a progressively smaller fraction of the gray areas it encloses.[21] Eventually, at very large scales, managerial knowledge becomes negligible compared to ignorance of what is being managed.

Managerial ignorance is unavoidable. As the size of an organization increases, it becomes progressively more important for the manager to recognize personal limits. At the limit of comprehension, the manager should stop pushing his or her physical boundary outward and delegate. At the organizational level, benign ignorance is certainly better than perverse ignorance, and it's often better than more knowledge. An effective organization can act coherently in the interest of its individual members precisely because its knowledge includes only the cooperative aspects of its members' knowledge. Effective members of an effective organization typically suppress and hide those aspects of their personal knowledge that do not contribute to organizational cooperation. And all members are better off. Petty bickering does not improve organizational effectiveness.

Cooperating with other individuals in the same system enhances individual fitness. Maintaining good relationships with the outside world also enhances individual fitness. Three of the five factors that Jared Diamond considers in *Collapse* (a discussion of the failure or success of human societies)[22] are at least partially outside-world factors—trading partners, hostile neighbors, and climate change. Because they are part of our ignorance, it's impossible to really understand such factors.

How should we behave in our relations with these unknown outside-world factors? They may be creatures like us. They may be creatures different from us. They may be organizations of creatures. They may be inanimate matter or forces. They may be combinations. Because we are humans, we think like humans, and that imposes an inherent limitation in our relationships with other creatures and other entities. The golden rule, "Do to others as you would have them do to you," is a reasonable rule for Homo sapiens, but it's certainly anthropocentric. It's also likely to be ethnocentric, and it may be centered on your own individual attitudes. Putting yourself in "the other person's shoes" is good in that it gives you a sense of another person's inherent value and potential capability, but it's dangerous if you try to use it to predict what another person will actually do. It becomes progressively more dangerous as the other person becomes less familiar or becomes a different kind of creature or an amorphous organization. It doesn't work at all for the details of an elemental force like a storm at sea.

So how should you deal with unavoidable ignorance of the outside world and unavoidable ignorance of whatever you're managing? The key is to maintain appropriate boundaries. Practice respect[23] for and be hospitable to the unknown, but also be wary of it. Don't do things that might create problems with the unknown. Be prepared for confrontations, but, when they come, give the benefit of the doubt before fighting. Never abuse the weak. Someday they may be stronger. Be courteous, friendly, helpful, kind, and generous, but avoid the presumption that you know better than others what's best for them. In other words, don't forcibly tear down their protective boundaries.

Feeling and showing a lack of respect for the unknown and being overly presumptive decrease the security of our systems, and that decreases our fitness. On the other hand, *if we limit our interactions with the unknown to what is necessary and helpful, and if we manage those interactions modestly, we'll enhance the strength of our systems, and that will enhance our fitness.*

CONCLUSION

Today's unconstrained flat world increases the number of options available to each locality, but it decreases local stability, it decreases the number of global options, it's very costly, and it's unsustainable. The flattening process stretches and weakens physical boundaries at all scales. This increases misalignment between physical boundaries and boundaries of consideration. And that increases perverse ignorance at all scales. The laws of thermodynamics and communications tell us that this cannot go on forever. Nature and culture will eventually react and reverse the process by erecting new physical boundaries. If we do not restrain ourselves, when the reversal occurs, other forces will impose restraints on us. *Our best strategy is to find appropriate places for our physical and intellectual boundaries and modulate those boundaries together in the gray area of uncertainty at the edges of our individual and collective comprehension.*

NOTES

1. In a separate essay in this volume, Wendell Berry provides an elegant taxonomy of the types of distortions and mistakes that can occur when boundaries are too distant and/or too weak.

2. Philip M. Morse, *Thermal Physics* (New York: Benjamin, 1965), chap. 17, "Entropy and Ensembles."

Information entropy (S) is determined as follows:

$$S = -\Sigma f_i \log_e(f_i),$$

where f_i = the probability of finding a particular set of information i when S is maximized subject to the constraints

$$\Sigma f_i = 1,$$
$$\Sigma f_i U_i = U,$$

in which U_i is the usefulness or "utility" of a particular set of information and U is average utility. Substituting the maximizing values of f_i back into the maximized equation and rearranging generates the emergent property,

$$T = (U - U_0)/S = \text{average net utility per unit of information.}$$

Thermodynamic entropy is the same, except S is multiplied by Boltzmann's constant, U is energy, and T is absolute temperature. Repeating the calculation for an ecological system with f_i = the probability of finding a particular species mix and U_i = abundance-weighted species rank in that mix makes T an intensive measure of diversity—a measure that is relatively independent of system size.

3. Simon Haykin, *Modern Filters* (New York: Macmillan, 1989), 42. The minimum sampling rate is called the Nyquist rate. The distortion is called *aliasing*. Aliasing is not random noise. It's an artifact of the relation between sampling and filtering, and it often looks like a legitimate signal.

4. Stuart A. Kauffman, *The Origins of Order: Self-Organization and Selection in Evolution* (Oxford: Oxford University Press, 1993); Per Bak, *How Nature Works: The Science of Self-Organized Criticality* (New York: Springer, 1996).

5. Heinz-Otto Peitgen, Hartmut Jürgens, and Dietmar Saupe, *Chaos and Fractals: New Frontiers of Science* (New York: Springer, 1992).

6. Seth Lloyd, "You Know Too Much," *Discover,* 28 April 2007, 53.

7. Thomas L. Friedman, *The World Is Flat* (New York: Farrar Straus Giroux, 2005).

8. We usually think of the nuclear envelop as protecting the chromosomes in the nucleus from disturbance by cell components in the cytosol outside the nucleus. But William Martin and Eugene Koonin ("Introns and the Origin of Nucleus-Cytosol Compartmentalization," *Nature* 440, no. 7080 [2 March 2006]: 41–45) suggest that the nuclear envelope may have evolved to provide protection in the other direction—to prevent the premature release of partially processed RNA molecules from the nucleus to the cytosol. Exposure to "half-baked" versions of such agents would wreak havoc in the cytosol and make the cell dysfunctional.

9. In *The Origin of Species* (1859), Charles Darwin summarizes this effect (in the final section of the second of two chapters entitled "Geographical Distribution") when he says: "We can see why whole groups of organisms . . . should be absent from oceanic islands, whilst the most isolated islands should possess their own peculiar species" (*The Origin of Species by Means of Natural Selection; or, The Preservation of Favored Races in the Struggle for Life* [New York: Modern Library, 1998], 548). In other words, the number of species on a small island is less than the number of species on a large continent. But the peculiarity of species on different islands makes the global number of species greater than if all the islands were connected in a single continent. For a quantitative description of this phenomenon, see Stephen Hubbell, *The Unified Neutral Theory of Biodiversity and Biogeography* (Princeton, NJ: Princeton University Press, 2001).

10. See, e.g., Herbert Goldstein, *Classical Mechanics* (Cambridge, MA: Addison-Wesley, 1950). A virial is the average excess (beyond-equilibrium) kinetic energy in a bounded system. When there is frictional dissipation, the virial can be expressed exclusively in terms of nonfrictional bounding forces and positions, but a separate flow of energy must be maintained to offset the dissipation.

11. *Strange images* is a metaphoric reference to the information "aliasing" described in n. 3 above.

12. Adding the time dimension creates a three-dimensional balloon whose volume has units of "action," which is energy × time. In classical mechanics (see Goldstein, *Classical Mechanics*), Hamilton's principle says that a dynamic system spontaneously moves through time in a way that minimizes total action—makes the three-dimensional balloon as small as possible. The Heisenberg uncertainty principle says that this total action can never be less than Planck's constant. But, at the human scale, sampling-rate limitations make the minimum total action many orders of magnitude greater. The limited sampling rate produces quantum mechanical distortion (perverse ignorance), but classical mechanics automatically filters it out.

13. Information is the time integral of information-flow rate. Knowledge is the time integral of information. If your boundary of consideration is less than your physical boundary, your knowledge is only part of the volume of your intellectual balloon. The rest of that volume is the time integral of aliased information—your perverse ignorance, or your energetic capability to make damaging mistakes.

14. Computer science pays close attention to an algorithm's "computational complexity"—the number of steps required to solve the problem. Typically, more efficient algorithms are harder to implement. For example, suppose you want to find the best of one thousand items. (*a*) In a "random" algorithm, you might randomly select one from the set and compare it with what you already have. If the new one is better, you keep it; otherwise you keep the one you already have. This is how evolution works. It's very simple, but it's slow. You need almost five thousand steps to find the highest value with more than about 99 percent confidence. (*b*) In a "greedy" algorithm, you might search the set sequentially. This guarantees finding the best value after exactly one thousand steps. (*c*) In a "cooperative" algorithm, you might set up what's called a heap—a binary tree in which each child is less than its parent. This always puts the highest value at the top of the heap, and you need only one step to find that highest value. But then you have a moral obligation to reconstruct the heap so that the next user can enjoy the same efficiency, and it takes about an additional ten steps to reconstruct the heap. That's still a good deal because eleven steps is almost one hundred times more efficient than one thousand steps, but we need a very stable system to enforce the moral obligation to reconstruct the heap after each user gets what he wants.

15. Dimitris Achlioptas, Assaf Naor, and Yuval Peres, "Rigorous Location of Phase Transitions in Hard Optimization Problems," *Nature* 435, no. 7043 (9 June 2005): 759–64.

16. These two attributes have analogous attributes in biology and physics. *Methods* (the number of ways of doing something) is analogous to biological diversity or physical pressure or temperature. It's an "intensive" variable. *Applications* (the number of replications) is analogous to the size of a biological population or physical area or volume. It's an "extensive" variable.

17. Harold J. Morowitz, *Foundations of Bioenergetics* (New York: Academic, 1978), 187.

18. Methods and applications can increase simultaneously, or one can increase before the other. In teaching or managing, stabilizing selection, or engine operation, the intensive variable increases first. In learning or inventing, speciation, or refrigeration, the extensive variable increases first.

19. Richard E. Michod, *Darwinian Dynamics: Evolutionary Transitions in Fitness and Individuality* (Princeton, NJ: Princeton University Press, 1999).

20. This statement is based on the species vs. area relationship proposed

by Olof Arrhenius ("Species and Area," *Journal of Ecology* 9, no. 1 [1921]: 95–99), which says:

(number of species in area 2)/(number of species in area 1) =

(area 2/area 1)^{Zspecies},

where $0.1 < Zspecies < 0.4$. Assuming a comparable energy/area vs. area relation, scaling up from the electronic energy and surface area of a hydrogen atom to the total biomass energy and surface area of the earth says that Zenergy/area = 0.16. Because the species in the species vs. area relation are separated, it's appropriate to divide Zenergy/area by $\log_e(2) = 0.693$, which suggests that, on average, Zspecies = 0.16/0.693 = 0.23, a value consistent with ecological measurements.

21. The white area and the gray areas it encloses are both fractals, but the white area's exponent is smaller. This is illustrated by the elegant flocking model in Iain D. Couzin, Jens Krause, Nigel R. Franks, and Simon A. Levin, "Effective Leadership and Decision-Making in Animal Groups on the Move," *Nature* 433, no. 7025 (3 February 2005): 513–16.

22. Jared Diamond, *Collapse: How Societies Choose to Fail or Succeed* (New York: Viking, 2005).

23. Paul Woodruff, *Reverence: Renewing a Forgotten Virtue* (Oxford: Oxford University Press, 2001).

TOWARD AN ECOLOGICAL CONVERSATION

Steve Talbott

The chickadee was oblivious to its surroundings and seemed almost ma-chinelike, if enfeebled, in its single-minded concentration: take a seed, deliver a few futile pecks, then drop it; take a seed, peck-peck-peck, drop it; take a seed . . . The little bird, with its unsightly, disheveled feathers, almost never managed to break open the shell before losing its talons' clumsy grip on the seed. I walked up to its feeder perch from behind and gently tweaked its tail feathers. It didn't notice.

My gesture was, I suppose, an insult, although I felt only pity for this creature—pity for the hopeless obsession driving it in its weakened state. There were several sick chickadees at my feeder that winter a few years ago, and I began to learn why some people view feeding stations themselves as an insult to nature. A feeder draws a dense, "unnatural" population of birds to a small area. This not only encourages the spread of disease but also evokes behavioral patterns one might never see in a less artificial habitat.

And, if feeders are problematic, what was I to think of my own habit of sitting outside for long periods and feeding birds from my hands? Especially during the coldest winter weather and heavy snow-falls, I sometimes found myself mobbed by a contentious crowd, which at different times included not only chickadees but also titmice, red- and white-breasted nuthatches, hairy woodpeckers, goldfinches, juncos, blue jays, cardinals, various sparrows, and a red-bellied woodpecker. To my great delight, several of the less wary species would perch on shoulders, shoes, knees, and hat as well as on hands.

But by what right do I encourage tameness in creatures of the wild?

The classic issue here has to do with how we should assess our impacts on nature. Two views, if we drive them to schematic extremes for purposes of argument, conveniently frame the debate.

On one side, with an eye to the devastation of ecosystems worldwide, we can simply try to rid nature of all human influence. The sole ideal is pristine, untouched wilderness. The human being, viewed as a kind of disease organism within the biosphere, should be quarantined as far as possible. Call this *radical preservationism.*

On the other side, impressed by our society's growing technical sophistication, we can urge the virtues of scientific management to counter the various ongoing threats to nature. Higher-yielding, genetically engineered vegetables, fruits, grains, livestock, fish, and trees—intensively monocropped and cultivated with industrial precision—can, we're told, supply human needs on reduced acreages, with less environmental impact. Cloning technologies may save endangered species or even bring back extinct ones. Clever chemical experimentation on the atmosphere could change the dynamic of global warming or ozone depletion.

Managerial strategies more appealing to many environmentalists include reintroduction of locally extinct species, collaring of wild animals for tracking and study, controlled predation by humans, and widespread use of bird nesting boxes—practices that have aided in the recovery of some threatened species, even if their lives must now follow altered patterns.

The problem with scientific management, founded as it is on the hope of successful prediction and control, is that complex natural systems have proved notoriously unpredictable and uncontrollable. Ecologists, writes Jack Turner in *The Abstract Wild,* keep "hanging on to the hope of better computer models and more information." But their hope is forlorn: "The 'preservation as management' tradition that began with [Aldo] Leopold is finished because there is little reason to trust the experts to make intelligent long-range decisions about nature. . . . If an ecosystem can't be known or controlled with scientific data, then why don't we simply can all the talk of ecosystem health and integrity and admit, honestly, that it's just public policy, not science?" "The limits of our knowledge," he adds, "should define the limits of our practice." We should refuse to mess with wilderness for the same reason we should refuse, beyond certain limits, to mess with the atom or the structure of DNA. "We are not that wise, nor can we be" (Turner 1996, 122–24).

Turner's critique of the ideal of scientific management is not unlike my own. But, as is usually the case with pitched battles between opposing camps, the real solution to the dispute between radical preservationists and scientific managers requires us to escape the assumptions common to both. Why, after all, does Turner agree with his opponents that acceptable "messing" with ecosystems would have to be grounded in successful prediction and control?

Once we make this assumption, of course, we are likely either to embrace such calculated control as a natural extension of our technical reach or to reject it as impossible. And yet, when I sit with the chickadees, messing with their habitat, it does not feel like an exercise in prediction and control. My aim is to get to know the birds and to understand them. Maybe this makes a difference.

It is certainly true, in one sense or another, that "the limits of our knowledge should define the limits of our practice." But we need to define the sense carefully. By what practice can we extend our knowledge if we may never act without already possessing perfect knowledge?

Our inescapable ignorance mandates great caution—a fact that our society has been reluctant to accept. Yet we cannot make any principle of caution absolute. The physician who construes the precept "First, do no harm" as an unambiguous and definitive rule can no longer act at all because only perfect prediction and control could guarantee the absence of harm. Those of us who urge precaution must not bow before the technological idols we are trying to smash. We can never perfectly know the consequences of our actions because we are not dealing with machines. We are called to live between knowledge and ignorance, and it is as dangerous to make ignorance the excuse for radical inaction as it is to found action on the boast of perfect knowledge.

There is an alternative to the ideal of prediction and control. It helps, in approaching it, to recognize the common ground beneath scientific managers and those who see all human "intrusion" as pernicious. Both camps regard nature as a world in which the human being cannot meaningfully participate. To the advocate of pristine wilderness untouched by human hands, nature presents itself as an inviolable and largely unknowable Other; to the would-be manager, nature is a collection of objects so disensouled and unrelated to us that we can take them as a mere challenge for our technological inventiveness. Both stances deprive us of any profound engagement with the world that nurtured us.

My own hope for the future lies in a third way. Perhaps we have missed this hope because it is too close to us. Each of us participates in at least one domain where we grant the autonomy and infinite worth of the Other while also acting boldly to affect and sometimes even rearrange the welfare of the Other. I mean the domain of human relations.

We do not view the sovereign individuality and inscrutability of our fellows as a reason to do nothing that affects them. But neither do we view them as mere objects for a technology of control.

How do we deal with them? We engage them in conversation.

WE CONVERSE TO BECOME OURSELVES

I would like to think that what all of us, preservationists and managers alike, are really trying to understand is how to conduct an ecological conversation. We cannot predict or control the exact course of a conversation, nor do we feel any such need—not, at least, if we are looking for a good conversation. Revelations and surprises lend our exchanges much of their savor. We don't want predictability; we want respect, meaning, and coherence. A satisfying conversation is neither rigidly programmed nor chaotic; somewhere between perfect order and total surprise we look for a creative tension, a progressive and mutual deepening of insight, a sense that we are getting somewhere worthwhile.

The movement is essential. This is why we find no conclusive resting place in Aldo Leopold's famous dictum: "A thing is right when it tends to preserve the integrity, stability, and beauty of the biotic community. It is wrong when it tends otherwise" (1970, 262).

Integrity and beauty, yes. But in what sense stability? Stability is not (as Leopold knew) mere stasis. Nature, like us, exists—preserves its integrity—only through continual self-transformation. Preservation alone would freeze all existence, denying the creative destruction, the urge toward self-transcendence, at the world's heart. Scientific management, on the other hand, reduces evolutionary change to arbitrariness by failing to respect the independent character of the Other, through which all integral change arises.

Turner, applying Leopold's rule to the past, is driven to suggest that "the last ten thousand years of history is simply evil" (1996, 35). He is, in context, defending the importance of moral judgment and passion. By all means, let us have moral indignation where it is due—and, heaven knows, plenty of it is due. But a ten-thousand-year history was simply evil? This is what happens when you make absolute a principle of stability and leave conversation and change out of the picture.

The antidote to Turner's stance here (a stance he himself continually rises above) is to consider what it might mean to engage nature in respectful conversation. One can venture a few reasonably straightforward observations.

In any conversation it is, in the first place, perfectly natural to remedy one's ignorance by putting cautious questions to the Other. Every experimental gardening technique, every new industrial process, every different kind of bird feeder is a question put to nature. And, precisely because of the ignorance we are trying to remedy, there is always the possibility that the question itself will prove indelicate or otherwise an occasion for trouble. (My bird feeder was the wrong kind, conducive to the spread of disease. And you can quite reasonably argue that I should have investigated the issues and risks more thoroughly before installing my first feeder.)

In a respectful conversation, such lapses are continually being committed and assimilated, becoming the foundation for a deeper, because more knowledgeable, respect. The very fact that we recognize ourselves as putting questions to nature rather than asserting brash control encourages us to anticipate the possible responses of the Other before we act and to be considerate of the actual response, adjusting ourselves to it when it comes.

This already touches on a second point: in a conversation, we are always compensating for past inadequacies. As every student of language knows, a later word can modify the meaning of earlier words. The past can, in this sense, be altered and redeemed. We all know the bitter experience of words blurted out unwisely and irretrievably, but we also know the healing effects of confession and penance.

This, in turn, points us to a crucial third truth. At any given stage of a conversation, there is never a single right or wrong response. We can legitimately take a conversation in any number of healthy directions, each with different shades of meaning and significance.

Moreover, coming up with my response is not a matter of choosing among a range of alternatives already there, already defined by the current state of the exchange. My responsibility is creative; what alternatives exist depends in part on what new alternatives I can bring into being. Gandhi engendered possibilities for nonviolent resistance that were not widely known before his time, and the developers of solar panels gave us new ways to heat our homes. If we have any "fixed" obligation, it is the obligation not to remain fixed but freely to transcend ourselves.

All conversation, then, is inventive, continually escaping its previous bounds. Unfortunately, our modern consciousness wants to hypostatize nature—to grasp clearly and unambiguously what this "thing" is so that we can preserve it. But the notorious difficulties in defining what nature is—what we need to preserve—are no accident. There is no such thing as a nature wholly independent of our various acts to preserve (or destroy) it. You cannot define any ecological context over against one of its creatures—least of all over against the human being. If it is true that the creature becomes what it is only by virtue of the context, it is also true that the context becomes what it is only by virtue of the creature.

This can be a hard truth for environmental activists to accept, campaigning as we usually are to save "it," whatever "it" may be. In conversational terms, the Other does not exist independently of the conversation. We cannot seek to preserve "it" because there is no "it" there; we can only seek to preserve the integrity and coherence of the conversation through which both it and we are continually transforming ourselves. Hypostatization is always an insult because it removes the Other from the conversation, making an object of it and denying the living, shape-changing, conversing power within it.

Finally, conversation is always particularizing. I cannot converse with an abstraction or a stereotype—a "Democrat" or a "Republican," an "industrialist" or an "activist," or, for that matter, a "preservationist" or a "scientific manager." I can converse only with a specific individual, who puts his own falsifying twist on every label I apply. Likewise, I cannot converse with a "wetland" or a "threatened species." I may indeed think about such abstractions, but this thinking is not a conversation, just as my discoursing on children is not a conversation with my son.

PERMISSION AND RESPONSIBILITY

How, then, shall we act? There will be many rules of thumb, useful in different circumstances. But I'm convinced that, under pressure of intense application, they will all converge on the most frightful, because most exalted, principle of all. It's a principle voiced, albeit with more than a little trepidation, by my Nature Institute colleague Craig Holdrege: "You can do anything as long as you take responsibility for it" (personal communication, 2001).

Frightful? Yes. The first thing to strike most hearers will be that impossibly permissive *anything*. What environmentalist would dare speak these words at a convention of American industrialists?

But hold on a minute. How could this principle sound so irresponsibly permissive when its whole point is to frame permission in terms of responsibility? Apparently, the idea of responsibility doesn't carry that much gravity for us—and isn't this precisely because we are accustomed to think of nature less in the context of responsible conversation than in that of technological manipulation? Must we yield in this to the mind-set of the managers?

If we do take our responsibility seriously, then we have to live with it. It means that a great deal depends on us—which also means that a great power of abuse rests on us. Holdrege's formulation gives us exactly what any sound principle must give us: the possibility of a catastrophic misreading in either of two opposite directions. We can accept the permission without the responsibility, or we can view the responsibility as denying us the permission. Both misreadings pronounce disaster. The only way to get at any balanced rule of behavior, any principle of organic wholeness, is to enter into conversation with it, preventing its diverse movements from running off in opposite directions, but allowing them to weave their dynamic and tensive unity through our own flexible thinking.

"You can do anything as long as you take responsibility for it." An ill-intentioned one-sidedness can certainly make of this a mere permission without responsibility. But, then, too, as we have seen, taking on the burden of responsibility without the permission ("First, do no harm—never, under any circumstance; do not even risk it") renders us catatonic.

Permission and responsibility must be allowed to play into each other. When we deny permission by being too assiduous in erecting barriers against irresponsibility, we are also erecting barriers against the exercise of responsibility. The first sin of the ecological thinker is to forget that there are no rigid opposites. There is no growth without decay and no decay without growth. So, too, there is no opportunity for responsible behavior without the risk of irresponsible behavior.

"But doesn't all this leave us dangerously rudderless, drifting on relativistic seas? Surely we need more than a general appeal to responsibility! How can we responsibly direct ourselves without an understanding of the world and without the guidelines provided by such an understanding?"

Yes, understanding is the key. We need the guidelines it can bring. But these must never be allowed to freeze our conversation. This is evident enough in all human intercourse. However profound my understanding of the other person, I must remain open to the possibilities of his (and my) further development—possibilities that our very conversation may serve. This doesn't, in healthy experience, produce disorientation or vertigo, a fact that testifies to a principle of dynamic (not static) integrity, an organic unity, at the foundation of our lives.

Such a principle, above all else, is what we must seek as we try to understand the world around us. The Nature Institute, where I work, sits amid the pastures of a biodynamic farm. The cows in these pastures have not been dehorned—a point of principle among biodynamic farmers. Recently, I asked Holdrege whether he thought one could responsibly dehorn cows, a nearly universal practice in American agriculture.

"How does dehorning look from the cow's perspective? That's the first thing you have to ask," he replied. When you observe the ruminants, he went on, you see that they all lack upper incisors and that they all possess horns or antlers, a four-chambered stomach, and cloven hooves. "If you look carefully at the animals, you begin to sense the significance of these linked elements even before you fully understand the relation between them. They seem to imply each other. Do you understand the nature of the implication? So here already an obligation presses upon you if you want to de-horn cattle: you must investigate how the horns relate to the entire organism." Given his own observations of the cow (Holdrege 1997), and given his discussions with farmers who have noted the different behavior of cows with and without horns, and given also the lack of any compelling reason for dehorning when the cows are raised in a healthy manner, Holdrege's own conclusion is: "Unusual situations aside, I don't see how we can responsibly dehorn cows."

Strange as this stance may seem outside a respectful, conversational context, it is a conclusion that remains natural to us at some half-submerged level of understanding. What artist would represent cattle without horns? (Picture the famous Wall Street bull dehorned!) The horns, we dimly sense, "belong" to these animals.

What the ecological conversation requires of us is to raise this dim sense, as best we can, to clear understanding. The question of what belongs to an animal or a plant or a habitat is precisely the question of wholeness and integrity. It is a question foreign and inaccessible to conventional thinking simply because we long ago quit asking it. We had to have quit asking it when we began feeding animal remains to herbivores such as cows and when we began raising chickens, with their beaks cut off, in telephone book–sized spaces.

Most dramatically, we had to have quit asking it by the time genetic engineers, borrowing from the philosophy of the assembly line, began treating organisms as arbitrary collections of interchangeable mechanisms. There is no conversing with a random assemblage of parts. So it is hardly surprising, even if morally debilitating, that the engineer is not required to live alongside the organisms whose destiny he casually scrambles. He is engaged not in a conversation but in a mad, free-associating soliloquy.

APPROACHING MYSTERY

Our refusal of the ecological conversation arises on two sides. We can, in the first place, abandon the conversation on the assumption that whatever speaks through the Other is wholly mysterious and beyond our ken. This all too easily becomes a positive embrace of ignorance.

I do not see how anyone can look with genuine openness at the surrounding world without a sense of mystery on every hand. Reverence toward this mystery is the prerequisite for all wise understanding. But *mysterious* does not mean "unapproachable." After thirty-two years of marriage, my wife remains a mystery to me—in some ways a deepening mystery. Yet she and I can still converse meaningfully, and every year we get to know each other better.

There is no such thing as absolute mystery. Nearly everything is unknown to us, but nothing is unknowable in principle. Nothing we could want to know refuses our conversational approach. A radically unknowable mystery would be completely invisible to us—so we couldn't recognize it as unknowable.

Moreover, the world itself is shouting the necessity of conversation at us. Our responsibility to avoid destroying the earth cannot be disentangled from our responsibility to sustain it. We cannot heal a landscape without a positive vision for what the landscape might become—which can only be something it has never been before. There is no escaping the expressive consequences of our lives.

Our first conversational task may be to acknowledge mystery, but, when you have prodded and provoked that mystery into threatening the whole planet with calamity, you had better hope that you can muster a few meaningful words in response, if only words of apology. And you had better seek at least enough understanding of what you have prodded and provoked to begin redirecting your steps in a more positive direction.

But claiming incomprehension of the speech of the Other is not the only way to stifle the ecological conversation. We can, from the side of conventional science, deny the existence of any speech to be understood. We can say, "There is no one there, no coherent unity in nature and its creatures of the sort one could speak with. Nature has no interior."

But this will not do either. To begin with, we ourselves belong to nature, and we certainly communicate with one another. So already we can hardly claim that nature lacks a speaking interior. (How easy it is to ignore this most salient of all salient facts!) Then, too, we have always communicated in diverse ways with various higher animals. If we have construed this as a monologue rather than a conversation, it is not because these animals offer us no response but only because we prefer to ignore their response.

But, beyond this, whenever we assume the organic unity of anything, we necessarily appeal to an immaterial "something" that informs its parts, which otherwise remain a mere disconnected aggregate. You may refer to this something as *spirit, archetype, idea, essence, the nature of the thing,* its *being,* the *cowness of the cow.* (Some of these terms work much better than others.) But without an interior and generative aspect—without something that speaks through the organism as a whole, something of which all the parts are a qualitative expression—you have no organism and no governing unity to talk about, let alone to converse with.

Remember: the science that denies an interior to nature is the same science that was finally driven by its own logic (e.g., in behaviorism) to deny the interior in man, a reductio ad absurdum if ever there was one. The same oversight accounts for both denials—namely, the neglect of qualities, which are the bearers of expression in both the world and the human being. Where there is genuine qualitative expression, something is expressing itself.

In his study of the sloth, Holdrege remarks that "every detail of the animal speaks 'sloth'" (1999, n.p.). Of course, you cannot force anyone to see the unity of the sloth—to see what speaks with a single voice (against standard evolutionary logic) through all the details—because you cannot force anyone to attend in a disciplined way to the qualitative substance of the world. But this much needs saying: a science that long ago decided to have nothing to do with qualities is not in a good position to tell those who do attend to qualities what they may or may not discover. (The stance of some churchmen toward Galileo's telescope comes to mind.)

What those who are receptive to the world's qualities consistently discover is a conversational partner.

WHERE DOES THE WILD LIVE?

To foreclose on the possibility of ecological conversation, whether owing to reticence in the presence of the mystery of the Other or simple denial of both mystery and Other, is to give up on the problem of nature's integrity and our responsibility. It is to forget that we ourselves stand within nature, bringing, like every creature, our own contributions to the ecology of the whole. Most distinctively, we bring the potentials of conscious understanding and the burden of moral responsibility. Can it be merely incidental that nature has begun to realize these potentials and to place this burden here, now, on us?

Raymond Dasmann sees wilderness areas as a refuge for "that last wild thing, the free human spirit" (quoted in Nash 2001, 262). The phrase is striking in its truth. But one needs to add that the human spirit is not merely one wild thing among others. It is, or can become, the spirit of every wild thing. It is where the animating spirit of every wild thing can be known, where it can rise to consciousness, where its interior speaking can re-sound under conditions of full self-awareness.

This is true only because, while we live in our environment, we are not wholly of it. We can detach ourselves from our surroundings and view them objectively. This is not a bad thing. What is disastrous is our failure to crown this achievement with the selfless, loving conversation that it makes possible. Only in encountering an Other separate from myself can I learn to love. The chickadee does not love its environment because it is—much more fully than we—an expression of that environment.

The willfulness and waywardness—the wildness—that has enabled us to stand apart and "conquer" nature is also what enables us to give nature a voice. The miracle of selflessness through which a human being today can begin learning to "speak for the environment"—a remarkable thing!—is the other face of our power to destroy the environment. So we now find ourselves actors in a grave and compelling drama rooted in the conflicting tendencies of our own nature, with the earth itself hanging in the balance. Given the undeniable facts of the situation, it would be rash to deny that this drama both expresses and places at risk the telos of the entire evolution of earth. But to accept the role we have been thrust into, and to sense our nearly hopeless inadequacy, is at the same time to open ourselves to the higher wisdom that would speak through us.

We do as much damage by denying our profound responsibilities toward nature as by directly abusing them. If you charge me with anthropocentrism, I accept the label, though on my own terms. If there is one creature that may not healthily scorn anthropocentrism, surely it is anthropos. How should we act, if not from our own center and from the deepest truth of our own being? But it is exactly this truth that opens us to the Other. We are the place within nature where willing openness to the Other becomes the necessary foundation of our own life.

The classicist Bruno Snell somewhere remarked that to experience a rock anthropomorphically is also to experience ourselves petromorphically—to discover what is rocklike within ourselves. It is the kind of discovery we have been making, aided by nature and the genius of language, for thousands of years. It is how we have come to know what we are—and what we are is (to use some old language) a microcosm of the macrocosm. Historically, we have drawn our consciousness of ourselves from the surrounding world, which is also to say that this world has awakened, or begun to awaken, within us (Barfield 1965, 1977).

In general, my observations of nature will prove valuable to the degree that I can, for example, balance my tendency to experience the chickadee anthropomorphically with an ability to experience myself "chickamorphically." In the moment of true understanding, those two experiences become one, reflecting the fact that my own interior and the world's interior are, in the end, one interior.

The well-intentioned exhortation to replace anthropocentrism with biocentrism, if pushed very far, becomes a curious contradiction. It appeals to the uniquely human—the detachment from our environment that allows us to try to see things from the Other's point of view—in order to deny any special place for humans within nature. We are asked to make a philosophical and moral principle of the idea that we do not differ decisively from other orders of life—but this formulation of principle is itself surely one decisive thing we cannot ask of those other orders.

There is no disgrace in referring to the "uniquely human." If we do not seek to understand every organism's unique way of being in the world, we exclude it from the ecological conversation. To exclude ourselves in this way reduces our words to gibberish because we do not speak from our own center.

But nothing here implies that humans possess greater "moral worth" (whatever that might mean) than other living things. What distinguishes us is not our moral worth but the fact that we bear the burden of moral responsibility. That this burden has risen to consciousness at one particular locus within nature is surely significant for the destiny of nature! When Jack Turner suggested that the last ten thousand years of human history may have been "simply evil," he ignored the worthy historical gift enabling him to pronounce such a judgment. How can we downplay our special gift of knowledge and responsibility without fatal consequences for the world?

TOWARD CREATIVE RESPONSIBILITY

We create "by the law in which we're made" (Tolkien 1947, 71–72). Our own creative speech is one, or potentially one, with the creative speech of nature that first uttered us. (How could it be otherwise?) This suggests that our relation to every wild thing is intimate indeed. We speak from the same source. We cannot know ourselves—cannot acquaint ourselves with the potentials of our own speaking—except by learning how those potentials have already found expression in the stunning diversity of nature.

Every created thing images some aspect of ourselves, an aspect that we can discover and vivify only through understanding. The destruction of a habitat and its inhabitants truly is a loss of part of ourselves, a kind of amnesia. Wendell Berry is right to ask: "How much can a mind diminish its culture, its community and its geography—how much topsoil, how many species can it lose—and still be a mind?" (Berry 2001, 50). As Gary Snyder puts it: "The nature in the mind is being logged and burned off" (quoted in Nash 2001, 263).

When Thoreau told us, "In wildness is the preservation of the world" (1947, 609), the wildness he referred to was at least in part our wildness. If humankind fails to embrace with its sympathies and understanding—which is to say, within our own being—every wild thing, then both we and the world will to that extent be diminished. This is true even if our refusal goes no further than the withdrawal from conversation.

Our failure to reckon adequately with the wild Other is as much a feature of human social relations as of our relations with nature and as much a feature of our treatment of domesticated landscapes as of wilderness areas. In its Otherness, the factory-farmed hog is no less a challenge to our sympathies and understanding than the salmon, the commonplace chickadee no less than the grizzly bear. We do not excel in the art of conversation. If the grizzly is absent from the distant mountains, perhaps it is partly because we have lost sight of, or even denigrated, the wild spirit in the chickadee outside our doors.

If we really believed in the saving grace of wildness, we would not automatically discount habitats bearing the marks of human engagement. We would not look down on the farmer whose love is the Other he meets in the soil and whose struggle is to draw out, in wisdom, the richness and productive potential of his farm habitat. Nor, thrilling to the discovery of a cougar track in the high Rockies, would we disparage the cultivated European landscape that, at its best, serves a far greater diversity of wild things than the primeval northern forest.

The point is not to pronounce any landscape good or bad but to ask after the integrity of the conversation it represents. None of us would want to see the entire world reduced to someone's notion of a garden, but neither would we want to see a world where no humans tended reverently to their surroundings (Suchantke 2001). We should not set the creativity of the true gardener against the creativity at work in our oversight of the Denali wilderness. They are two very different conversations, and both ought to be—can be—worthy expressions of the wild spirit.

A WORD UNASKED FOR

In late winter or early spring, the chickadee flock frequenting my feeder begins to break up as the birds pair off for mating. Only two (with their offspring) will occupy a given territory, and during summer those few may rarely visit a feeder; there are too many superior delicacies around.

A few summers ago, I decided not to maintain a feeder and, because of other preoccupations, scarcely noticed any chickadees on the property. They were the furthest thing from my mind when, on a warm August day at a time of extraordinary personal distress, I happened to be standing outside in a small clearing. There was no brush or other bird cover immediately at hand. Suddenly, a chickadee came out of nowhere and alighted on the fence railing four or five paces in front of me.

Standing still, I watched for several seconds as it regarded me with an apparently intense interest. Then, instead of veering away as I expected, it flew with its soft, stutter-step flight straight toward me, dipping characteristically a few inches in front of my nose before rising as if to land on my bald pate. But, with a slight hesitation, it seemed to have second thoughts (there's not much of a perch up there) and passed on behind me. This unlooked-for gesture from a "long-lost friend"—a moment of mutual recognition recalling an earlier conversation—touched me deeply. In the flush of affection I felt for the creature granting me this unexpected interview, I found an easing of my pain. Its life was so free, so far removed from my own problems, yet it was so precious . . .

"That's very nice, but do you really glorify this encounter as part of a meaningful conversation? And do you believe that the chickadee was responding to your inner condition at the time?"

Well, hardly. I am serious—and I include myself in the rearmost rank—when I say we have scarcely learned to converse with nature (or, for that matter, with each other). But, nevertheless, one can at least glimpse the beginnings of conversation here.

The very first—and, perhaps, the most important—conversational step we can take may be to acknowledge how we have so far failed to assume a respectful conversational stance. For example, how much of my activity in feeding the birds by hand is driven by my self-centered pleasure in their attentions rather than by selfless interest in who they are and what they need? To ask such a question is already to have shifted from manipulator to listener.

But, no, I would not claim that the chickadee on the fence railing was sympathizing with my troubles. Of course, because of my ignorance, and because the chickadee is a speaking presence, I cannot say absolutely that it was not, at some level of its being, responding to my inner condition or that it was not the agent of some sort of Jungian "synchronicity." But I am skeptical, and such things are in any case wholly beyond my knowledge. So I leave them alone.

What I do know is that the chickadee was, in an obvious and un-problematic sense, responding to me in its expressive, chickadee-like manner. And this manner was partly familiar to me because I have paid attention to the chickadees in my neighborhood. The behavior, even if unexpected, was not altogether strange to me. I could say, "Yes, if a chickadee were to gesture in my direction, that is how it might do it; it was just like a chickadee." And, in saying this, I could bring to mind much about the chickadee's way of speaking itself into the world. This, in turn, gives me something to respond to, something to respect, something to make a proper place for both in the world and in myself.

And, yes, maybe even something to invite in certain directions through attentive, reverential conversation. I do still occasionally feed the birds from my hands. This is a behavior they would never engage in if there were no humans in the world, but I have yet to see that it in any way diminishes them. I am more inclined to think the opposite. Chicka-dees are known to have a great curiosity about other creatures, along with a particular affinity for humans, and giving a few of them a little room to explore this affinity does not seem such a bad thing.

There are, of course, appropriate limits. Personally, I draw the line when the chickadees try to use my mustache as nesting material.

REFERENCES

Barfield, Owen. 1965. *Saving the Appearances.* New York: Harcourt, Brace, Jovanovich. Originally published in 1957.
———. 1977. *The Rediscovery of Meaning, and Other Essays.* Middletown, CT: Wesleyan University Press.
Berry, Wendell. 2001. *Life Is a Miracle: An Essay against Modern Supersti-tion.* Washington, DC: Counterpoint. Originally published in 2000.
Holdrege, Craig. 1997. "The Cow: Organism or Bioreactor?" *Orion,* Winter, 28–32.
———. 1999. "What Does It Mean to Be a Sloth?" *NetFuture,* no. 97. Avail-able at http://netfuture.org/1999/Nov0399_97.html#2.
Leopold, Aldo. 1970. *A Sand County Almanac: With Essays on Conservation from Round River.* New York: Ballantine.
Nash, Roderick Frazier. 2001. *Wilderness and the American Mind.* New Ha-ven, CT: Yale University Press.
Suchantke, Andreas. 2001. *Eco-Geography: What We See When We Look at Landscapes.* Great Barrington, MA: Lindisfarne.

Thoreau, Henry David. 1947. "Walking." In *The Portable Thoreau,* ed. Carl Bode, 592–630. New York: Viking. Originally published in 1862.

Tolkien, J. R. R. 1947. "On Fairy Stories." In *Essays Presented to Charles Williams.* Oxford: Oxford University Press.

Turner, Jack. 1996. *The Abstract Wild.* Tucson: University of Arizona Press.

IGNORANCE AND ETHICS

Anna L. Peterson

In his essay in this volume, Wes Jackson challenges us to take ignorance seriously in politics, science, and a host of other fields. Since we are so much more ignorant than knowing, he asks, why not go with our strong suit? This question opens up a host of intriguing possibilities for a variety of disciplines and fields. I concentrate here on the implications for social ethics, beginning with two questions. First, do our ideas about doing good and being a good person depend on knowledge we have or think we have? And, second, if we acknowledge that we have much less knowledge than we think—and that the sum total of what we know will always be much smaller, and much less certain, than the sum of what we do not know—then what does this mean for thinking and acting ethically? After exploring the role of knowledge in several ethical systems, I look at some of the possible models for an ignorance-based ethic. The most systematic of these are religious in origin, although some secular alternatives also exist. Their "ignorance" is most evident in their visions of human nature and of historical change, in which admissions of ignorance ground a willingness to relinquish not only knowledge claims but also means-ends calculations.

CRITIQUING KNOWLEDGE-BASED ETHICS

When we scratch the surface, most of the ways we usually think about doing and being good turn out to be based on knowledge we have or think we have. Without the right knowledge, we assume, it is impossible to act morally. This knowledge might be quite general, such as "knowing" about human nature. If we know what people are really like, for example, we can provide fitting rewards or punishments and, thus, encourage morally good behavior. Or we can predict how someone will act in a given situation, which is important if we seek a particular morally desirable result or conclusion.

Moral behavior may also require more specific kinds of knowledge, such as the appropriate rules to follow in a given situation, the wishes of others, or the likely risks of a certain action. Doing good thus depends on answers to questions such as the following: Do we have all the relevant historical and empirical information? Do we know about all the interested parties? How accurately can we assess present conditions and predict future ones? Without access to this sort of knowledge, our efforts to do the right thing seem doomed to failure. How can I decide what to do in a given situation without knowing how others will act or what may be the result of my own actions? Does the likelihood of success justify my effort? And what do I risk in my effort to do good?

The questions asked by professional philosophers are not all that different, nor are their assumptions and conclusions about the relation between knowledge and morality. This becomes clearer if we look at two of the most important models in philosophical ethics. The first, consequentialist ethics, are those in which the goodness of an action or a decision is based on its outcome. Results matter much more than processes, in this approach, though extreme versions in which ends justify all means are rare, at least philosophically. The best-known consequentialist philosophy is utilitarianism, including the influential models developed by John Stuart Mill and Jeremy Bentham. According to utilitarian thought, the moral course is that which ensures "the greatest possible good for the greatest possible number." There are many varieties of utilitarianism, but they share a common base in certain kinds of knowledge. Utilitarians assume that we "know" that all people are self-interested, that all want to attain similar goods and to avoid similar evils, and, perhaps most important, that particular decisions or acts will have certain consequences. Lack of correct knowledge, in this perspective, is likely to lead to bad consequences and, thus, to bad morals.

A second important framework for philosophical ethics emphasizes rules or laws. The classic example of this approach is Immanuel Kant's "categorical imperative," according to which people should act according to principles ("maxims," for Kant) that they believe should be universal laws. For Kant, a rule or principle that is moral must by definition be universalizable and not dependent on any particular context or characteristics of the subject. An example of this is human rights discourse, according to which disregard for context, consequences, and personality is precisely what defines morality. For this sort of ethics, the most important kind of knowledge involves how to frame the rules or maxims that should direct moral behavior. For a maxim to be universalizable, it must be phrased with great care, as are Kant's various formulations of the categorical imperative.

Unlike utilitarianism and other consequentialist models, rule-based ethics do not require information about consequences. The hope or likelihood of a particular result should not influence adherence to rules. Thus, for example, a rule-based ethic might assert that the torture of prisoners is always wrong, even if it may generate vital information, or that telling the truth is always right, even if it will hurt someone's feelings. Nor do rule-based ethics require knowledge about the details of the situation since, once the right principle is found, it should be generally applicable. Being generally applicable, in fact, is what defines it as "moral." The main kind of knowledge needed for ethics in this model is how to frame the appropriate moral laws. For some sets of rules, however, further knowledge concerning the feasibility of a moral action is necessary. Kant asserted that we are not obligated to adhere to principles that cannot be carried out: "ought," as he put it, implies "can." This guideline seems to make sense—but only if we can know, with reasonable certainty, what actions and outcomes are both possible and likely.

Consequentialist and rule-based ethics are often seen as polar opposites: the first makes ethics dependent on the outcome of an action, while the second judges morality according to preestablished principles and deliberately ignores consequences. Despite these differences, the two approaches share a dependence on knowledge claims. In this respect, they have common ground with a number of religious ethics. However, for religious ethics, the knowledge required for morally good behavior encompasses not only human nature, moral rules, or the likely outcomes of certain actions but also the sacred. This is especially clear in the "ethical monotheism" of Judaism, Christianity, and Islam. Here, the link between knowledge and morality is crucial because correct behavior is generally seen as a result of correct belief, which, in turn, is based on correct knowledge—or, rather, on what religious people claim to know.

Here we encounter a difficult issue: secular philosophers, and indeed many religious believers, would agree that the "knowledge" that people can have of the sacred, for example, of the existence of God or of the exploits of Krishna, is rarely amenable to empirical or historical verification. It is faith, and not knowledge, that undergirds most religious ethics, and, while faith may seek understanding, as Saint Anselm put it, understanding or knowledge is rarely the starting point for faith. That said, most religious people do indeed claim to "know" certain things about the nature of the sacred, the acts or teachings of divine figures, the origins and destiny of the created world, and a host of other issues that shape moral behavior. In this sense, many religious ethics are indeed knowledge based. I know that Jesus died for my sins, a believer might contend, or that the Four Noble Truths are indeed true, or that Muhammad is indeed the last true prophet, so I know what I must do to be a good person. There are, I will argue later, some religious ethics that do not rely on knowledge in this sense, just as there are some secular moralities that do not depend on knowledge claims. If a religious believer's model for ethical behavior depends on what she or he claims to know, however, the fact that this "knowledge" strikes others as unverified, perhaps even obviously false, does not make it an ignorance-based ethic. What is important, at least for my purposes here, is that the claimed knowledge justifies and guides the believer's ethic, not that the knowledge has been subjected to empirical verification.

Within this broad definition, a host of otherwise diverse religious ethics can be described as knowledge based. Within Christianity, for example, the dominant Roman Catholic approach based on Thomas Aquinas's natural law theory presumes that humans can, for the most part, know what God wants because the human conscience contains innate knowledge of God's laws. Furthermore, not only can people know the divine will, but they can also act in accord with it, aided by conscience and practical reason. Aquinas was supremely optimistic about knowledge: through the human faculties of conscience, reason, and established laws, he thought, natural law allows us to acquire knowledge of divine law and apply its principles in concrete cases.[1] These assumptions about the relation between human reason, knowledge, and morality continue to shape modern Roman Catholic thought.

Protestantism rejected natural law theory as both arrogant and misguided. Martin Luther insisted that the gap separating humanity and God in both knowledge and morality was so wide that mortals could never know God's will or act in accord with it. In Luther's view, natural law theory is presumptuous to the point of sacrilege in its assumption that humans can reason out what God wills. While historical Protestant-ism thus appears to reject knowledge as a basis for morality, a closer look reveals that knowledge remains central to the ethics of Luther and other Protestant founders, including John Calvin. They do indeed dis-miss the capacity of human reason to know how the world works as well as the ultimate moral value of any knowledge gained by purely human means. However, mainstream Protestant faith turns out to be based not on an embrace of ignorance but, rather, on a different sort of knowledge—that granted by faith in Christ. To "know Jesus"—and to know him in a particular way, mediated by particular assumptions and objectives—is the key both to salvation and to moral life in the world.

Despite real differences, dominant Catholic and Protestant ap-proaches agree that ethical acts, as well as ultimate salvation, require certain kinds of, and certainty about, human knowledge. This does not, however, appear to be such a problem. Why should we not base moral actions on knowledge? Why would anyone want a morality based on ignorance, anyway? I argue that basing morality on knowledge claims generates a host of political and indeed ethical problems, including de-nial of moral diversity, the premature closure of possibilities, and over-simplification of both the moral issues at stake and the consequences of any given act. Basing ethics on ignorance may not resolve all these difficulties, but uncovering the claims underlying knowledge-based eth-ics and exploring some of their drawbacks can encourage necessary reflections on both the assumptions behind and the consequences of principles and values that we often take for granted.

One problem with knowledge-based ethics is that they provide an easy way to dismiss challenges as insufficiently or inaccurately in-formed. This is the voice of "If only they knew what we knew, they would come around to our way of valuing things." This presumes that some sort of moral progression is possible, with increasing levels of knowledge as the stepping stones, and that the culmination of this as-cent is our own ethical perspectives and values. This approach denies the possibility that smart, knowledgeable, and well-meaning people might have different values from our own. It denies, in other words, the reality of moral diversity. This denial points to the fact that the knowl-edge underlying "knowledge-based" ethics is not just about specific content, such as what Kant, the law, or a sacred scripture says. More often than not, the really important ethical knowledge is less obvious and more general; it is about how people and the world "really" are. If we "know" that people are self-interested, for example, then a whole host of claims fall into place. We know what people will do in a given situation, and we act accordingly. This view of morality requires denying our profound and unavoidable ignorance about what people do and want in a given situation.

This is linked to another problem with knowledge-based moralities, which is the way they close off not only possibilities of human diversity but also human and nonhuman unpredictability. Humans, other animals, the weather, the whole natural world, and even human inventions can all surprise us. They do not always do what we think they will do (or want them to do). This is part of the deep ignorance of life on this planet, for humans and all creatures. Incorporating the possibility and even the likelihood of surprise, and of error, into our ethics is more realistic than basic moral evaluations, decisions, and actions based on assumptions of correct knowledge. It is more than likely that our assumptions and predictions are wrong, in other words, and our ethics will be better if we take seriously the likelihood of error from the very beginning.

An example illustrates some of the problems with knowledge-based ethics. One of the most influential moral approaches to violence is just-war theory. Traditionally, theories of just war enumerate a series of conditions, categorized under two broad headings. The first involves the justice of the war (*jus ad bellum*). It asks whether going to war is a last resort after all peaceful means to end a conflict have been exhausted, whether the cause is just, whether the leader calling for war is a legitimate authority, and whether one's side is likely to prevail. The second set of rules concerns the justice with which the war is conducted (*jus in bello*). It asks, for example, whether the violence is proportionate to the provocation and whether prisoners and civilians are treated fairly. If all the conditions (which vary according to the model adopted) cannot be met, then a nation or group may not justifiably take up arms. Just-war theory has developed as part of Western Christian social thought, with strong influences from Augustine and Thomas Aquinas, and is the dominant approach for evaluating armed conflicts among Christian churches today. The alternative, adopted by a small minority of Christians, is pacifism, which asserts that organized violence against other people is never justified under any conditions.

Just-war theory represents an excellent example of a knowledge-based ethic, and a closer look at this model sheds light on some failings of knowledge-based ethics more generally. While just-war models vary among themselves, most share several sorts of knowledge claims. First, they usually assume knowledge about what might or might not have worked—explicit in the claim that we can know when "all peaceful means have been exhausted." Even more strikingly, just-war theories usually insist on a good likelihood of success as a prerequisite for going to war. All the just causes in the world cannot justify taking up arms, in other words, if your side appears doomed to failure. This links power to moral right. It also, not incidentally, ignores the fact that sometimes David beats Goliath.

Jean Bethke Elshtain, a prominent contemporary just-war theorist, provides an excellent example of the pitfalls of knowledge-based ethics. Elshtain is unusually explicit about the dependence of her ethical model on knowledge claims or, as she puts it, on "getting the facts straight."[2] She labels those with whom she disagrees—pacifists, opponents of war—ignorant. They do not know how the world really works; Elshtain does. She assumes, further, superior knowledge not only about human nature and reality itself but also about the likely results of particular actions, such as military interventions. Elshtain runs into problems, however, and these underline some of the pitfalls of knowledge-based ethics more generally. What she claims as accurate knowledge often turns out to lack support—for example, her description of government knowledge and actions prior to the invasion of Iraq. Because she has insisted that the authority of her ethic is its foundation in "fact," when the facts prove wanting, her ethic crumbles.

For knowledge-based ethics, whose primary claim to moral authenticity is superior knowledge, any incorrect or incomplete knowledge undermines the ethical system. No room is left for surprise, historical openness, human error, or nature's unpredictability. Knowledge turns out to be a weak ground for moral authority—credibility vanishes the first time the facts turn out to be not as advertised. More important than credibility, perhaps, is the incapacity to motivate and sustain ethical action. If the motivation for "doing good" rests on knowledge that turns out to be shaky, the temptation may be strong to withdraw into apathy or simply to pursue self-interest. This knowledge-based ethic reinforces many destructive features of our culture, including a desire for control over inherently unstable and diverse realities, a drive to oversimplify, and blindness to the concerns and priorities of others. The ethics of knowing that dominate politics and public discourse in this country often help bolster the status quo and those in power. They fail, in turn, to nurture qualities we desperately need, including humility, curiosity, and the courage to take risks.

IGNORANCE-BASED ETHICS

Ignorance does not mean the rejection of all knowledge. Rather, it entails an acknowledgment of how much we do *not* know, coupled with an awareness that anything we claim to know, perhaps especially about human and nonhuman nature, we know only partially and tentatively, and this is always subject to revision. (This does not imply, it is worth pointing out, that an ignorance-based worldview, and its corresponding ethic, must go as far as some postmodernist approaches in denying the possibility of *any* knowledge.) We need to approach knowledge, and action that leads from knowledge, with humility because we are fallible. This position is not simply a form of ethical relativism, according to which all knowledge and all actions are morally equal, because we have no way of judging the very claim of equality. We have some knowledge, but it is limited, changeable, and, above all, partial. We can hold better and worse knowledge but never any complete, perfect, or final knowledge. And, whatever we know, it is always from a specific, partial perspective—a standpoint that fundamentally shapes our way of knowing as well as the content of our knowledge.[3]

An ignorance-based ethic not only acknowledges uncertainties but places them at the heart of its own justifications and moral claims. This need not mean embracing stubbornness or intolerance. Instead, it entails questioning all claims to complete, objective knowledge. Taking absolute correct knowledge as impossible, an ignorance-based ethic takes as its very foundation an expectation of mistakes and surprises. This differs sharply from the various ethical models that rest, in the end, on a conviction that humans are at core knowing creatures and that it is precisely their ways of knowing, and the fruit of that knowing, that can, should, and do ground their moral decisions and acts. Ignorance here resembles what Paul Tillich calls the "Protestant Principle," a prophetic protest against every power that claims divine character for itself.[4] Claims to ultimacy, for Tillich, lead to idolatry, which "sets up something finite as absolute."[5] This, I think, points to the deepest problem with knowledge-based ethics. Claims to know certain things are a poor foundation for visions of the good, in no small part because they tend to increase both our good opinion of ourselves and the urgency with which we cling to these opinions. If, as frequently happens, our knowledge turns out to have been incomplete or simply wrong, we are forced either to construct a new ethical compass or—more often—to find new "knowledgeable" justifications to continue as before (the "rubber ruler"). What is at stake here is not a particular item that is known but, rather, the claim to be knowledgeable, to be a particular kind of creature and to relate to the world in a particular way, from a vantage point that sets us apart from other creatures and from the situations in which we are immersed.

A good model of a religious ignorance-based ethic lies in the Radical Reformation or Anabaptist movement, which began shortly after Martin Luther launched the Protestant Reformation in 1517. Central to Anabaptist faith is a conviction that Christians are called to build the reign of God on earth, in and through a church reflecting the values of this reign. This effort demands separation from the inevitable corruption, conflict, and violence of the secular realm. This commitment to the "kingdom as social ethic," in the words of the Mennonite theologian John Howard Yoder,[6] may be Anabaptism's defining trait and what most clearly separates it from mainstream Protestantism. From the beginning, Anabaptists have challenged establishment Christianity for its compromises with the sword, its complacent acceptance of the world's sins, its willingness to settle for "civic peace" and relegate the reign of God to the dustbin of theology. The Christian duty, Anabaptists insist, is not to make pacts with the powers that be but, rather, to live as though the reign of God were indeed in their midst while admitting that, ultimately, this reign is not a human project. Anabaptist faith is not a millenarian hope for the future realization of the reign of God but, rather, an attempt, inevitably partial and flawed, to create it here and now. The fact that this Christian community may always be a minority, even a remnant, makes its obligations no less binding.

Anabaptist morality makes a particular kind of ignorance foundational to definitions of goodness and principles for action. It recognizes, first, that certainty is always elusive and that no person is omniscient. Further, this ignorance-based model of ethics not only acknowledges its lack of certain pieces of knowledge but insists that full knowledge is both impossible and unnecessary. We can and indeed must find grounds for acting morally without full knowledge of humans, the world, or God. Grounds for moral action—for discipleship—stem, rather, from a hopeful faith that God exists, is good, and calls people to live in a society of peaceableness and solidarity. Living in the way God calls people to live demands not only faith but also a relinquishment of knowledge, both about what people are like and about what might happen in a given situation. Anabaptist ethics thus challenge traditional Christian formulations of the relation between belief and knowledge. They are neither "understanding seeking faith" nor "faith seeking understanding." These two traditional approaches invert each other's ordering but still posit fundamental links between knowledge and a faith-based ethics.

In the Radical Reformation model, in contrast, although both understanding and ethics have a place, they remain separate: morality does not depend on understanding or knowledge. Anabaptists believe that God calls Christians to build the kingdom on earth while simultaneously asserting that God provides no certainty that this task can be achieved. This is the deep ignorance at the heart of their ethic: believers must live as if the reign of God were a real human society, as if a real human society could be the reign of God, without knowing whether that is possible, and without any possibility of acquiring that knowledge. Knowledge that humans could build the reign of God would entail a loss of submission to God; knowledge that humans could *not* build the kingdom of God would entail a loss of faith in God. And both faith and submission are central to Anabaptist identity and life. What holds the two together is an admission of profound and permanent ignorance.

Anabaptists are not the only Christians to have based their ethics on an admission of ignorance. Christian existentialism, for example, agrees that knowledge and rationality are insufficient to explain the world or to guide human action in it. For Søren Kierkegaard, this meant that the Christian, or the "knight of faith," must live by "leaps" made with faith but without reasonable certainty or proof. Paul Tillich updated this theology in *The Courage to Be,* in which he argued that the courage necessary to persevere can be sustained only if it is grounded in something greater than nonbeing. For Tillich, as for Kierkegaard, this "ground of being" cannot be rationally known. It is possible only as a result of faith, which Tillich defines as the "existential acceptance of something transcending ordinary experience."[7] This Christian existentialism is not irrational: reason has its place, but it also, perhaps more important, has its limits. Insofar as Christian existentialism acknowledges the impossibility of full understanding or knowledge, it resembles Anabaptist faith.

In ignorance-based ethics such as the Anabaptist model, right action tends to be more important than right knowledge. Anabaptists not only affirm that "faith without works is dead" (from the Letter of James, which Luther despised) but also believe that value can be found in good works undertaken without right belief or knowledge. This suggests that ignorance-based ethics entail a new approach not just to knowledge but to action. Rather than the idea—common to mainline Protestant, natural law, and Kantian ethics—that right knowledge must come first, ignorance-based ethics ask how good action can occur despite the lack of full and adequate knowledge. Given that our ignorance will always outweigh our knowledge, how might we discern what is good and how to do it?

On this issue, Buddhism offers some ignorance-based insights. In Buddhism, right action can and often should precede right knowledge. A Buddhist might undertake practices, such as meditation, without a full spiritual or philosophical understanding of their meaning. In and through practice, in fact, greater knowledge and understanding should emerge. There is no need to wait for certain knowledge to begin action, and, in fact, knowledge is generally not possible apart from practice. Understanding can follow action, in contrast to both mainstream Protestantism and Enlightenment philosophy, both of which make intentionality and "the will" central to determining the moral value of an action.

It is no accident that Buddhist and especially Anabaptist ethics have adopted pacificism as their ethical-political response to most historical situations. While I do not propose an exact overlap between ignorance-based ethics and pacificism, I believe that the application of ignorance-based ethics to actual human conflicts will lead to pacificist conclusions more often than not. Pacifism is almost always a more humble and cautious approach than the use of force. As I noted in relation to just-war theory, the use of force by nations presumes knowledge both about future events and about human nature. Pacifists usually remain agnostic on these issues, preferring to leave as many possibilities open as possible. Violence, especially violence wielded by institutions such as a national government, leaves little room for uncertainty. The decision to attack an enemy or to initiate a war presupposes confident knowledge on a number of fronts, including the rightness of the attacker's cause, the attacker's ability to prevail, and the enemy's response. An ignorance-based worldview admits that it is impossible to have sure knowledge about most or any of these factors. An ignorance-based ethic would add that the costs of being wrong in a war are so great as to be avoided at almost any cost.

The examples of ignorance-based ethics above all have religious foundations, which raises the question of whether "deep ignorance" provides a ground of being for secular ethics. What might prove an adequate foundation for a secular ignorance-based ethic? Aristotelian virtue theory offers one possible example, particularly in its insistence that moral judgment need not wait for full knowledge and in the admission that ethics always entails uncertainty. These principles emerge in Alasdair MacIntyre's influential updating of virtue theory, *After Virtue*. It is crucial, MacIntyre asserts, "that at any given point in an enacted dramatic narrative we do not know what will happen next." Unpredictability and teleology, MacIntyre suggests, are held together in human life. We do not know what will happen next, but our lives have a certain form that projects itself toward the future. "There are constraints on how the story can continue and . . . within those constraints there are indefinitely many ways that it can continue."[8] MacIntyre critiques Marx, in fact, for presenting the narrative of human life as law governed and predictable in a certain way.

On the other hand, MacIntyre appreciates the "deep optimism" of Marxism. This optimism, I suggest, is at odds with Marx's claims to know how history will proceed—his unwillingness to acknowledge unpredictability, as MacIntyre points out. The claim to know what will happen in the future is not easily compatible with optimism, I argue, not because optimists do not often think they know what will happen, but because they are so often proved wrong—and then disillusioned. Acknowledging ignorance about the future (among other things) provides a more solid ground for enduring hopefulness.

MacIntyre also shares with Marx a critique of the "detached self" of liberalism. This individualistic self contrasts with the "narrative self" that MacIntyre proposes, for which "the story of my life is always embedded in the story of those communities from which I derive my identity. I am born with a past; and to try to cut myself off from that past, in the individualist mode, is to deform my present relationships." What I am, he summarizes, is not just what I choose to be but also what I inherit.[9] This social view of the self assumes a high degree of ignorance since, when you throw in your lot with others and admit that you are mostly not the master of your own fate, then you admit that you cannot know many things, including what others think and feel, their pasts, their passions, where they might head.

An ethic grounded on knowledge that is partial, incomplete, and subject to change could be guided by humility and cooperation rather than arrogance and domination. In such a framework, moral decisions would begin not with certainty about our knowledge but, rather, with an acknowledgment that what we know is provisional and that we can never anticipate all the consequences of our actions. We must "learn to live fully and act creatively in the midst of a world we can never control and can only partially understand," as the feminist ethicist Sharon Welch proposes.[10] This sort of ethic, Welch suggests, might resemble jazz improvisation, which fundamentally relinquishes several kinds of knowledge: about how things will turn out, how another person will respond to an action, even how we will feel after each step.

Similar questions are asked by those feminist ethicists who complement, or in some cases replace, rational principles and knowledge with "care." Rather than centering ethics on abstract and absolute principles such as rights, care ethicists turn to relationships and feeling. This alternative emerges from the work of scholars such as Carol Gilligan, an educational psychologist whose 1982 book, *In a Different Voice,* argued that women and men make moral decisions in very different ways. Boys tend to view ethics in relation to abstract principles, while girls more often base moral decisions on their relationships.[11] Gilligan terms these the *justice* and *care* perspectives: "From a justice perspective, the self as moral agent stands as the figure against a ground of social relationships, judging the conflicting claims of self and others against a standard of equality or equal respect. . . . From a care perspective, the relationship becomes the figure, defining self and others." These different emphases lead to distinctive moral questions and responses. In the care perspective, Gilligan proposes, the question at stake is not "What is just?" but, rather, "How to respond?"[12] These questions suggest the different degrees of certainty involved in the two approaches. The question of what is just demands a definitive answer based on confident knowledge, while asking how to respond presumes a more tentative and partial approach, one that presumes not certainty as much as compassion and the willingness to go forward in a caring manner.

Care ethics do not overlap entirely with an ignorance-based approach. Because care ethics are personal, concrete, and contextual, they require detailed knowledge of the social and emotional context for a particular moral decision, in contrast to the more abstract, formal, and general approach of justice models. The knowledge demanded by care ethics must rest on careful attention and dialogue and is limited by the histories, needs, and interests of the parties involved. Humility and openness are required by the ongoing dialogues and mutually determining experiences that are integral to care ethics. This contrasts with both the abstractness and the definitiveness, the "once and for all" character, of most knowledge-based ethics.

Feminist models such as those proposed by Welch and Gilligan, among others, resolve a number of the intellectual and moral problems that arise when ethics are tied to knowledge. However, a crucial dilemma remains: What motivates action in the absence of certainty? An endless improvisation may not keep all people going, especially in the face of risk, inconvenience, disappointment, and other challenges. As Paul Tillich contended, mere open-endedness may not be sufficient: the courage to act ethically requires a foundation in something greater than the nonbeing that threatens. This does not require knowledge about other people, the results of our actions, or even our own capacities. It does not even require knowledge of the "something greater" that grounds our work and lives. We admit that we can never fully know or explain the ground that makes possible our being and acting, which always exceeds our grasp. Still, such a ground is necessary, and faith in this ground is required for action of any sort, from risky activism to simply getting through the challenges of everyday life.

CONCLUDING REFLECTIONS

The most important contribution of an ethic based on ignorance, or, more precisely, on a frank admission of ignorance, is to provide grounds for hope and a spur to action, which are vanishingly hard to find in these cynical and pessimistic times. It might seem paradoxical to suggest that embracing the permanent inadequacy of our knowledge can provide hope, but I think that it can. The acknowledgment that we can never be sure about what we know or what will happen next frees us from the cost-benefit calculation that dominates so many styles of moral decisionmaking. Whether a desired result is likely no longer becomes a major motivating factor in moral decisions. This is important because, as Yoder argues, illusions of certain knowledge often tempt people to calculate likely causes and effects, which, in turn, often lead to exhaustion or abandonment of a cause when calculations prove inaccurate.[13]

In an ignorance-based model, we no longer assume that we can know what people will do and what will result from the actions of various parties, including ourselves. In the absence of such knowledge, moral decisions and actions are grounded in relationships and in people's individual and collective capacities to persevere in the face of uncertainty. These relations provide one possible ground for an ignorant (and hopeful) ethic. Another possibility is found, for many people, in nonhuman nature. The "more than human" world is as unpredictable and unmanageable, and just as clearly the source of life and power, as any deity. Some make their "nature religion" explicit, while for others the spiritual dimensions of their relationship to nature remain unspoken, sometimes unrecognized. But this spiritual dimension is not hard to uncover: like the reign of God, nature remains always beyond our grasp, our knowing, and our planning. Nature is a force, or set of forces, that grounds ignorance just as does the finally unknowable God of existentialist and Anabaptist Christianity.

Founding our moral acts on ignorance can remind us that the future is open, despite what "common sense" (or the powers that be) claims to know. Tillich addresses this in his nuanced argument for a critical utopianism as ground for hope. This utopianism rests on an admission of ultimate ignorance, combined with an unwavering commitment to continue regardless of knowledge or outcome. Confronting the rise of Nazism in his native Germany, Tillich wrote: "We face here the problem of not using the forces of fanaticism and yet of demanding an unconditional commitment against them in the hour of necessity. In committing ourselves, however, we know that we are not committed to something absolute but to something provisional and ambiguous and that it is not to be worshiped but criticized and if necessary, rejected; but in the moment of action we are able to say a total Yes to it."[14]

This "yes," for Tillich, is a form of "anticipation" that acknowledges ultimate ignorance about that which is believed in. "The thing ultimately referred to in all genuine anticipation remains transcendent; it transcends any concrete fulfillment of human destiny; it transcends the otherworldly utopias of religious fantasy as well as the this-worldly utopias of secular speculation." And, we might add, it transcends all certain knowledge. We can never bring about, Tillich writes, a situation that is exempt from the permanent uncertainties of human existence, but this "does not mean that distorted reality should be left unchanged."[15] We do not know what will happen; we are always, necessarily ignorant about the consequences and the ultimate context of our acts; and this is never justification for failure to act.

NOTES

1. Thomas Aquinas, *Summa Theologica,* in *Thomas Aquinas on Law, Morality, and Politics,* ed. William P. Baumgarth and Richard J. Regan (Indianapolis: Hackett, 1988), esp. question 91. The one form of law that is beyond human knowledge is that granted through revelation, embodied in sacred scripture, and even there Thomas sees no fundamental contradiction between human rationality and divine law.

2. Jean Bethke Elshtain, *Just War against Terror* (New York: Basic, 2003), 14, 15.

3. For more on this view of knowledge, see the work of feminist theorists, such as Donna Haraway's *Simians, Cyborgs, and Women: The Reinvention of Nature* (New York: Routledge, 1991).

4. Paul Tillich, *The Protestant Era* (Chicago: University of Chicago Press, 1957), 230.

5. Paul Tillich, *Political Expectation* (New York: Harper & Row, 1971), 177.

6. John Howard Yoder, *The Priestly Kingdom: Social Ethics as Gospel* (Notre Dame, IN: University of Notre Dame Press, 1984), chap. 4, "The Kingdom as Social Ethic."

7. Paul Tillich, *The Courage to Be* (New Haven, CT: Yale University Press, 1952), 173.

8. Alasdair MacIntyre, *After Virtue: A Study in Moral Theory* (Notre Dame, IN: University of Notre Dame Press, 1981), 200, 201.

9. Ibid., 205, 206, 244.

10. Sharon Welch, *Sweet Dreams in America: Making Ethics and Spirituality Work* (New York: Routledge, 1999), 51.

11. Carol Gilligan, *In a Different Voice: Psychological Theory and Women's Development* (Cambridge, MA: Harvard University Press, 1982), 33.

12. Carol Gilligan, "Moral Orientation and Moral Development," in *Women and Moral Theory,* ed. Eve Feder Kittay and Diana T. Meyers (New York: Rowman & Littlefield, 1987), 23.

13. Yoder, *Priestly Kingdom,* 97.

14. Tillich, *Political Expectation,* 178.

15. Tillich, *Protestant Era,* 172.

IMPOSED IGNORANCE AND HUMBLE IGNORANCE—TWO WORLDVIEWS

Paul G. Heltne

There are at least two kinds of ignorance in the human world. These differ so dramatically that they may be seen as opposing worldviews. One ignorance worldview is built on the belief that one knows or understands a situation or a subject rather thoroughly, perhaps even definitively or absolutely, when, in fact, one does not. This sort of ignorance-masquerading-as-certain-knowledge often comes to us as whole systems of thought and work and with intellectual buffers that make its facts, claims, and practices beyond question. Its assumptions, often invisible or unstated, are thereby unassailable. You know that you are in the presence of this kind of ignorance when you are made to feel that it would be foolish to ask questions about derived conclusions or about basic assumptions. This kind of ignorance is, thus, purposely imposed on many and camouflages our true state of ignorance.

Wes Jackson has demonstrated that the metatechnology of modern agriculture is just such a purposeful ignorance; indeed, Jackson may be the first person ever to have articulated the basic assumptions of agriculture and held them up to examination. That does not keep modern agriculture from being a very cherished and comfortable pattern of thinking and acting for lots of folks, on the farm and off. Since we all cherish, in some fashion, what we are used to and do not like to be embarrassed, purposeful ignorance can be long-lived indeed.

There is another kind of ignorance. Acknowledging that one does not know is a humble kind of ignorance, one that is, in fact, filled often with the joy of discovery and wonder at what is discovered. This is the kind of ignorance-based worldview that can help us fathom the messes we are in, articulate assumptions and processes, entertain questions and be enriched by them, and imagine new ways and new knowledge.

Humble ignorance can imagine that it might be wrong and hopes that its community will correct it early enough to avoid harm. It can marvel at what it sees that it cannot hope to understand or control. It knows that it must question certainty and jargon.

I believe that purposeful ignorance is deeply rooted in the process of our cultural systems and possibly in the processes of our consciousness. It may seem contradictory, but the patterns of purposeful ignorance often include scientific ways of thinking and speaking. The propensity toward purposeful ignorance may deepen as scientific insight moves toward management applications. Below, I offer two brief illustrations to show how deeply ingrained and potent purposeful ignorance is. In contrast, I also propound an alternative way of looking at nature that may help keep us in the humility of an ignorance-based worldview.

MICROFAUNA

By way of an initial illustration, let me quote a brief passage from an Aldo Leopold article, "The Last Stand," written sixty-five years ago. Leopold describes how a Swiss forest has been yielding high-quality timber since the seventeenth century under a regime known as selective harvesting. A contiguous forest area of the same kind of timber was clear-cut in the seventeenth century and has never recovered despite intensive forestry care. Leopold then notes that:

> Despite the rigid protection, the old slashing now produces only mediocre pine, while the unslashed portion grows the finest cabinet oak in the world; one of those oaks fetches a higher price than a whole acre of the old slashings. On the old slashings the litter accumulates without rotting, stumps and limbs disappear slowly, natural reproduction is slow. On the unslashed portion litter disappears as it falls, stumps and limbs rot at once, natural reproduction is automatic. Foresters attribute the inferior performance of the old slashing to its depleted microflora, meaning that underground community of bacteria, molds, fungi, insects, and burrowing mammals which constitute half the environment of a tree.

Now this is the part that applies to ignorance; Leopold continues: "The existence of the term microflora implies, to the layman, that science knows all the citizens of the underground community, and is able to push them around at will. As a matter of fact, science knows little more than that the community exists, and that it is important. In a few simple communities like alfalfa, science knows how to add certain bacteria to make the plants grow. In a complex forest, science knows only that it is best to let well enough alone."[1] In the same article, Leopold lamented the rapid clear-cut logging of the Porcupine Forest, the last remaining sizable stand of old-growth forest on the Upper Peninsula of Michigan. Indeed, the Swiss example was given to show that there were other good, sustainable ways of managing a forest. Leopold decried the fact that in 1942 in the United States we continued clear-cutting even though the alternative of selective logging was well and widely understood. The greedy arrogance of the forestry industry regularly destroyed the very conditions of growing timber for the future.

There is a terrible tragedy of purposeful ignorance in the unquestioned forestry and economic rationales driving that kind of destruction. That greedy foolishness and perverse refusal to cap one's greed had in it an expression of purposeful blindness. Grim, indeed, is the fact that, to this day, we are still clear-cutting the forests of the United States. Is there an old-growth forest that our forestry industry would not like to cut?

But there is another expression of purposeful ignorance that Leopold brings to our attention when he unmasks the word *microflora*. In fact, *microflora* conceals vast ignorance with the appearance of knowledge—and not just to the layman. The word can, as Leopold suggests, allow scientists, too—particularly those who are not experts on the particular microflora—to think that a great deal is known, that the complexity of the community is fathomed, and that, as the microflora is apparently understood, parts of the community might be substitutable, unnecessary, or that the community might function apart from the shelter and sustenance of the mature forest. Thus, we may lose the idea that the myriad, mutual adaptations of forest and soil are, in fact, astonishing and a thing to be marveled at. Without an admission of our ignorance, we may be unable to perceive that the complex forest community possesses its own biotic rights (to borrow another phrase from Leopold).[2]

By contrast, purposeful ignorance has clearly not led to humility in the face of the forest's complexity or to a celebration of the sustaining nature of the flourishing mutual interactions of the trees and soil that we may never fully understand. Leopold's example shows us that we may, however, choose a humility-based worldview that starts from the point of admitted ignorance and leads us to continuous renewal and celebration.

BIODIVERSITY

Biodiversity and *microflora* share characteristics. Biodiversity can be a chilly, counting way of describing some parts of nature. Primack defines the concept of *biological diversity* (the term he nearly always uses rather than *biodiversity*) as including genetic variation, species diversity, and community diversity.[3] But, of course, it is not possible to measure or describe all these things at once by means of a single number. In conservation biology and ecology, biodiversity is usually a count or sum total of counts by which the number of life-forms in one region or community can be compared with those in another. Biodiversity counts describe a current state of existence, almost a snapshot in time. The counts have no ability to convey information about the actual condition of the community (except that we have an inherent sense that more is better) or the historical trends in the community's condition.

That biodiversity can be a troubled metric is shown by a simple thought experiment. Suppose that, in a particular county, a fine-textured-soil savanna community might occur in only one or two of the many sites where it originally occurred—in other words, the community is found in a few isolated locations. In some kinds of descriptive accounts, the countywide biodiversity figure for that community might be reported as the same whether the occurrences were few or many. Even if the actual number of communities were reported, we would not thereby know whether that number was wonderful or perilous or whether the fine-textured-soil savanna community had once covered the whole county. Thus, it is important to note that, by itself, even on an ecosystem scale, a biodiversity count might only faintly suggest the vibrant interactions of living communities of plants and animals that are still so little understood. As with the word *microflora,* the word *biodiversity* can suggest that we know much more than we do about the life of natural communities.

Thus, I think that biodiversity can easily slide into the realm of purposeful ignorance, where it can have many serious corollaries. For instance, since biodiversity reports a count of something and, in sum, purports to provide a kind of "bottom-line" snapshot of nature in a region, economists seem somewhat able to relate to nature described in this way. With nature reduced to the numbers with which they are comfortable, the economists then are emboldened to ask that we attach a monetary value to species, communities, and ecosystem services that would justify the public expense needed to conserve the biodiversity. With this question, the frame of reference of the conservation discussion switches dramatically *from* nature with its own biotic rights and well-being *to* valuations based on assigned dollar amounts (and, of course, we humans assign the figures—by actually asking people what they would be willing to pay for a species, ecosystem, etc.). A large part of the conservation community seems to have bought into this economic reframing of the argument, perhaps because the argument takes up the biodiversity metric or because they have become convinced that the monetary argument is the only one that sways the minds of public officials. But, of course, the fact is that we are far too ignorant about the ecosphere or biotic communities to think that we could frame arguments about them using the reductive assignment of a dollar value. Once begun on this pathway, it is easy to end up by talking as though all the "services" coming out of nature are available for use by humans—a serious and destructive ignorance.

Fortunately, the larger public seems to have a much humbler sense of its relationship to nature, a sense that is not reduced to dollar values but, rather, is expressed in terms of concerns and commitments. As reported by the *Biodiversity Recovery Plan,* prepared by the Chicago Regional Biodiversity Initiative, the Chicago-area public has a view of nature characterized by "feelings of ethical obligations to protect species from extinction, religious values associated with cherishing the Earth and its inhabitants, and the desire to leave for future generations that which we are able to enjoy." The *Plan* then goes on to quote from a biodiversity report as follows: "In some ways, these concerns are the core motives for protecting biodiversity. A national survey of public attitudes about biodiversity, a survey that included focus groups in Chicago, found that responsibility to future generations and a belief that nature is God's creation were the two most common reasons people cited for caring about conservation of biodiversity."[4] This finding seems to me to reveal a kind of humble sensibleness that recognizes responsibility toward a future of which we are and always will be largely ignorant. Furthermore, persons with this kind of humble ignorance have voted overwhelmingly for bond issues to protect and restore nature in the region. These contrasting approaches lead one to ask, Is the commitment of conservation biologists to the term *biodiversity* getting in the way of saving biodiversity? How would the approach to the public differ if professional conservationists operated primarily from a language of *values* that was frankly based on ecological humility and wonder, as contrasted to their current reliance on monetary *valuations,* the economistic approach?

AN ALTERNATIVE WAY OF LOOKING AT THINGS

I have used Aldo Leopold's discussion of microflora to alert us to the fact that a word (and especially a scientific term) can mask a great deal of complexity and a great deal of ignorance about that complexity. Subsuming complexity in glib simplicity almost forces the audience into ignorance and, thus, may clear the way for destructive practices against nature. For example, we clearly know how to manage forests in a way that keeps the complex forest-soil community intact. However, promoting the forest community as a simple system—the kind of imposed ignorance elided by presumptions attached to words like *microflora*—certainly helps cut back on the hard questions that should be advanced about forestry decisions. As a further example, the oversimplifications associated with the term *biodiversity* are in contrast to the thoughts and actions of those who see nature as a whole, who perceive a nature uncounted and undiscounted, who simply don't allow the nature they know to be spoken of in reductionistic terms.

To help us in this battle against invalid reductionism, I offer an image that keeps complexity before the mind's eye. The polytope in figure 1 is such an image. For scientist and layperson alike, such an image can provide a reminder of our ignorance and an ongoing admission of it. This image is a defense against imposed ignorance posing as certainty while glossing over the complex interconnected being of nature.

For some time now I have been deeply intrigued with the idea of many-to-many connectivity because this is what I think goes on in nature, as contrasted with one-to-one relationships. Figure 1 is an example of a complex regular polytope. A polytope is used here as a way of imaging, by an esthetically pleasing analogy, multiple interconnections across dimensions and scales. Figure 1 is a specific multidimensional polytope projected onto two dimensions. This is an area of geometry developed by H. S. M. Coxeter at the University of Toronto.[5] I suggest that figure 1 presents an image analogy for envisioning the true

Figure 1. A complex regular polytope. From H. S. M. Coxeter, *Regular Complex Polytopes,* 2nd ed. (Cambridge: Cambridge University Press, 1991), 44. (Courtesy of the Department of Mathematics, University of Toronto. See further http://www.math.toronto.edu/gif/polytope.gif.)

complexity of the workings of a cell, an organism, a plant and animal community, an economy, or even perhaps the functioning of the World Wide Web.

So many attempts to depict the astonishing connectivity that occurs in living systems (especially in textbooks) are pale shadows of the systems' complexity. However, I point to several other images that are specific depictions of interconnections. Figure 2 is the partial protein interaction map of a *Drosophila melanogaster* cell. It is partial because proteins with more than twenty interactions are omitted from the figure; including them would make it too dense to convey meaning. This depiction is remarkable for many reasons. For instance, for some proteins to "interact" requires the product of one reaction to get to another part

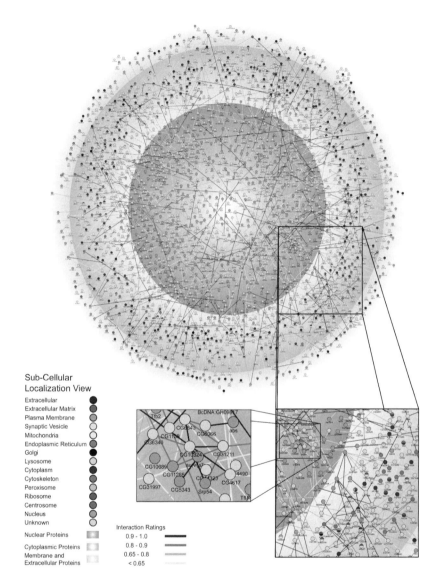

Figure 2. Global views of the protein-interaction map. Filtered by only show-ing proteins with less than or equal to twenty interactions. This results in a map with 2,346 proteins and 2,268 interactions. (From L. Giot et al., "A Protein Interaction Map of *Drosophila melanogaster,*" *Science* 302, no. 5651 [2003]: 1727–36, fig. 4. Reprinted with permission of the American Association for the Advancement of Science.)

of the cell, or even from outside the cell membrane all the way into the nucleus, for the next part of the anabolic or catabolic process. Imagining how these interconnections and processes work, and, indeed, learning about them, leads one to astonishment.

Figure 3 depicts the food webs for tropical fish communities in several rain forest streams in Venezuela. These webs show complex, many-to-many interactions. Some ecologists would suggest that the food webs depict most of the important interactions in a community because they show who eats whom and who may be in competition with whom. However, I submit that, as complex as these food webs are, the total set of interactions is much more diverse than this in nature and, thus, that any depiction of the whole set of community interactions is correspondingly more complex. Figure 4 shows that even the simplified complexity of food web interactions varies over time. This figure depicts the food webs in two small English streams, in March and again in October.

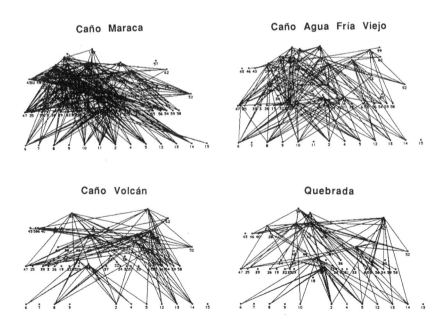

Figure 3. Community food webs in several Venezuelan and Costa Rican tropical streams, omitting weak trophic links. (From K. O. Winemiller, "Spatial and Temporal Variation in Tropical Fish Trophic Networks," *Ecological Monographs* 60, no. 3 [September 1990]: 331–67.)

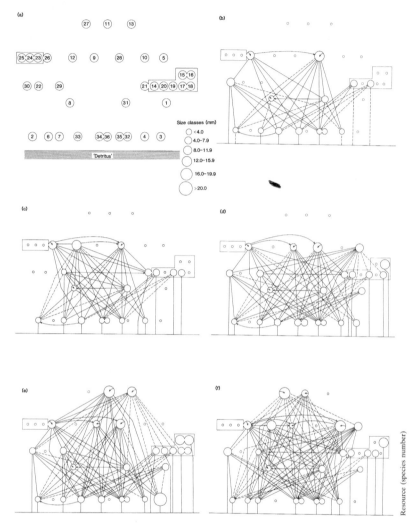

Figure 4. Food web diagrams for selected streams in England: (*c*) open water, June; (*d*) open water, October; (*e*) margin, March; (*f*) margin, June. (From P. H. Warren, "Spatial and Temporal Variation in the Structure of a Fresh Water Food Web," *Oikos* 55 [1989]: 299–311.)

CLOSING REMARKS

Many other examples of the profound complexity of nature could be given. But, in conclusion, I want to return to the image of the complex, regular polytope (fig. 1 above) and imagine how it might function as a depiction of a natural community in a particular place. (I specifically do not use Golley's word *ecotope* because, as Golley defines it, an ecotope is a place on a landscape small enough that the "abiotic patterns are defined as relatively uniform."[6] However, uniformity is dependent on the scale of the biota being considered, and discussion of biotic community engages in imposed ignorance when it cannot envision trees and microscopic organisms engaged in interaction, especially as the abiotic surroundings of the tree would probably not be uniform with respect to the microbiota.)

Why ought we to attend to the polytope analogy? To see why, it may be useful to bring the complex, regular polytope from an analogical realm into what might be called a dynamic image of the complex, irregular world. For the sake of discussion, let us say that the nodes of the polytope represent species and the lines represent the flow of matter, energy, or information. Consequently, to deepen our analogy, we need to imagine that all the lines are not going to be equally bright all the time. Some interconnections may function only rarely but could be of ultimate importance, such as sporadic predation. In the winter season or the dry season, the lines representing some interactions might fade out for long periods. Some of the nodes could represent species that migrate and join the community for only a few weeks or months of the year—nevertheless, those weeks could be crucial for the migrant species as well as for the transfer of energy, material, and information in the whole system for the remainder of the year. Even when the connections are established, the content of the connections will vary from day to night or from moment to moment. The nodes will have varied connections when their populations are diseased or well; when they are cold or warm, wet or in drought; when they are young, mature, or dead; or in the presence of toxins or predators. The polytope thus becomes this glimmering, pulsating entity—beyond even the powers of complex regular polytopes to describe adequately (H. S. M. Coxeter, personal communication). Of course, the whole multidimensional vessel is afloat on the environmental sea, and each individual in each of its nodes is affected differentially by that milieu.

Fundamentally, I suggest the polytope image because it is an apt way to keep ever present in our minds an idea of the complex arrangements of which we are a part and on which we depend for everything that makes life possible. It is a visual reminder that insistently positions discoveries made on single dimensions of the natural world within the larger question, What don't we understand? Thus, the polytope image is, I think, a safeguard against certain kinds of imposed ignorance. If terms like *microflora* could be printed with an asterisk in the image of a polytope, it might be much harder to risk placing the microfauna under any threat. *Biodiversity* followed by such an asterisk might persistently bring to mind the vision of a vibrant, complex biotic community. Perhaps the polytope asterisk would help us relate to diversity in such a way that we would find it harder to think that we could pay for it with dollars and make it easier for others to share in our wonder and concern for natural communities.

Indeed, nature is a home of grace and beauty of which we are largely ignorant and yet the home that we must care for. We are just coming to realize the existence of this degree of complexity at all levels of biological existence; we have little ability to understand such complexity, yet we are charged to protect it undiminished for all those who come after us. This beautiful image of a polytope presents a useful reminder of our humble ignorance in the face of such complexity. The polytope provides a vision that encourages us to recognize the parts of our complex relationship while keeping the incomprehensible whole in the mind's eye. We need such an ever-present reminder to increase the likelihood that we will keep nature entire for the future.

NOTES

I wish to thank Brooke Hecht and Anja Claus of the Center for Humans and Nature and the reviewers and editors for their thoughtful advice.

1. Aldo Leopold, "The Last Stand," *Outdoor America* 7, no. 7 (May–June 1942): 8–9, reprinted in *The River of the Mother of God and Other Essays by Aldo Leopold,* ed. Susan L. Flader and J. Baird Callicott (Madison: University of Wisconsin Press, 1991), 290–94.

2. Aldo Leopold, *A Sand County Almanac and Sketches Here and There* (Oxford: Oxford University Press, 1949), 211.

3. His definition (from Richard B. Primack, *Conservation Biology,* 2nd ed. [Sunderland, MA: Sinauer Associates, 1998], 23) reads as follows:

The protection of biological diversity is central to conservation biology, but the phrase *biological diversity* can have different meanings. The World Wildlife Fund . . . defines it as "the millions of plants, animals, and microorganisms, the genes they contain, and the intricate ecosystems they help build into the living environment." By this definition, biological diversity must be considered on three levels:

> *a*) *Species diversity.* All the species on Earth, including bacteria and protists as well as the species of the multicellular kingdoms (plants, fungi, and animals);
>
> *b*) *Genetic diversity.* The genetic variation within species, both among geographically separated populations and among individuals within single populations;
>
> *c*) *Community diversity.* The different biological communities and their associations with the physical environment ("the ecosystem").

4. Chicago Regional Biodiversity Council, *Biodiversity Recovery Plan* (Chicago: Chicago Regional Biodiversity Council, 1999), 14.

5. See H. S. M. Coxeter, *Regular Complex Polytopes,* 2nd ed. (Cambridge: Cambridge University Press, 1991).

6. F. B. Golley, *A Primer for Environmental Literacy* (New Haven, CT: Yale University Press, 1998), 99.

PART THREE

Precursors and Exemplars

BATTLE FOR THE SOUL OF IGNORANCE

Rhetoric and Philosophy in Classical Athens

Charles Marsh

For nobility and pathos, not many moments in the history of philosophy rival the apologia and death of Socrates.

Calm, deliberate, Socrates stood before his accusers and—he who incessantly interrupted opponents throughout the Platonic dialogues—begged not to be interrupted.

But his defense failed, as he knew it would. He wouldn't play to the jury, scorning "the artificial language of a schoolboy orator" (Plato 1989a, 17c [*Apology*]). The jury's sentence, we know, was death. The charges, Plato tells us, were corrupting youth and worshiping false gods.

Of those charges, we might say, paraphrasing Professor Higgins of *My Fair Lady:* "How tragic! How heartrending! How . . . accurate."

Accurate? Socrates corrupting youth and worshiping false gods?

Such a contrarian charge requires an explanation. The clarification will lead us from Socrates to Plato and then to Aristotle and Isocrates, all of them struggling in the dawn of philosophy to define the soul of ignorance and how it should function in civilized society.

PLATO'S SOCRATES AND THE SOUL OF IGNORANCE

The explanation begins with Chaerephon—"You know Chaerephon, of course," Socrates tells his accusers (Plato 1989a, 21a [*Apology*]). A boyhood friend of Socrates', Chaerephon had long ago sought the oracle of Delphi and asked whether anyone were wiser than Socrates. And the oracle had replied no. No one was wiser than Socrates.

Confessing that he had no claim at all to wisdom at the time of the oracle's assertion, Socrates tested the oracle by visiting wise, accomplished Athenians. He began with politicians, then moved to poets and skilled craftsmen. With each alleged wise man, he came to the same conclusion: "He thinks that he knows something which he does not know, whereas I am quite conscious of my ignorance." And from this repeated discovery he discerned the true meaning of the oracle's utterance: "The wisest of you men is he who has realized, like Socrates, that in respect of wisdom he is really worthless" (Plato 1989a, 21d, 23b [*Apology*]).

Here, surely, is the ideal platform for the adoption of ignorance as a worldview, as a framework for social debate. Why not nominate Socrates as the poster boy of a humble, deferential approach to the gathering of knowledge? Unfortunately, Socrates—at least the Socrates in Plato's dialogues—didn't stop there. Socrates believed that ignorance was curable. He believed that in its soul, so to speak, ignorance was finite and conquerable.

According to Plato's Socrates, philosophers were the conquerors of ignorance. In using this word—*philosopher*—Plato expanded its meaning beyond, simply, "a lover of wisdom." "Philosophers are those who are capable of apprehending that which is eternal and unchanging," he writes in the *Republic*. "Those who . . . wander amid the multiplicities of multifarious things are not philosophers" (1989c, 484b). In *The Open Society and Its Enemies*, Karl Popper (1966), no fan of Plato's, writes: "Plato gives the term *philosopher* a new meaning, that of . . . a seer of the divine world" (145)—of God's mind, in short. To add spice to this notion of philosophy—to move philosophy beyond some esoteric backroom debate—we should note, as does Harvey Yunis (2003) in *Written Texts and the Rise of Literate Culture in Ancient Greece,* that in classical Athens "science and philosophy . . . constitute a single intellectual enterprise" (12). To cut to the chase, Plato's Socrates is telling us that wise men can discover absolute, unchallengeable scientific truth.

In the best Socratic tradition, we should challenge this new definition and ability of philosophy and science. "So, Socrates," we should ask, in the question-and-answer wrangle that he loved, "how can philosophers reach the mind of God and find absolute truth?" Socrates' pupil Phaedrus asks much the same thing as he and Socrates stroll outside the walls of Athens. As the teacher and student recline in the shade of a plane tree, Socrates may injure his reputation with some of us by declaring: "I'm a lover of learning, and trees and open country won't teach me anything, whereas men in the town do" (Plato 1989b, 230d [*Phaedrus*]). But Phaedrus lets this pass unchallenged, and he eventually asks how true wisdom is acquired. Socrates answers by discussing "the dialectic method," a "discourse" that involves "the processes of division and bringing together, as aids to speech and thought" (Plato 1928, 276e, 266b [*Phaedrus*]). Philosophers engaged in dialectical conversation, Socrates maintains, define, analyze, synthesize, repeat the cycle—and arrive, eventually, at the wellspring of knowledge and reality: an idea in the mind of God. The dialectic of Plato's Socrates is a kind of exclusive Jacob's Ladder—philosophers only, please—that takes qualified seekers from the earth to heaven and back again.

Plato believed that all aspects of earthly reality were corrupt forms of original, divine ideas. Says Socrates in the *Republic:* "Perhaps there is a pattern . . . laid up in heaven for him who wishes to contemplate it." And indeed there is, even for things as mundane as household furniture: "Now God, whether because he so willed or because some compulsion was laid upon him not to make more than one couch in nature, so wrought and created one only, the couch which really and in itself is. But two or more such were never created by God and never will come into being. . . . If he should make only two, there would again appear one of which they both would possess the form or idea, and that would be the couch that really is in and of itself, and not the other two." Dialectical philosophers thus reason their way "up to the first principle itself" (Plato 1989c, 592b, 597c, 533c), which will seem strangely and wonderfully familiar because our souls experienced it with God in heaven before we were born into earthly existence. In debunking this intellectual homecoming in his *History of Western Philosophy,* Bertrand Russell (1945) writes that such moments of piercing, emotional insight are actually ephemeral parts of the creative process: "Every one who has done any kind of creative work has experienced, in a greater or less degree, the state of mind in which, after long labour, truth, or beauty, appears, or seems to appear, in a sudden glory—it may be only about some small matter, or it may be about the universe. The experience is, at the moment, very convincing; doubt may come later" (123).

But for the Platonic philosopher, doubt doesn't come later—in fact, if doubt appears in others, the Platonic philosopher must crush it. This intolerance dramatically increases the seriousness of the battle for the soul of ignorance. It's one thing for Plato's Socrates to say that philosophers know what's what, that they've reasoned their way onto the celestial couch. But it's another thing entirely for philosophers to aggressively and unfairly quell any dissent. Yet that's exactly what Socrates recommends as he establishes a connection between philosophy and rhetoric, the art of persuasion. In the *Gorgias,* Socrates disparages rhetoric—which is based only on opinion, he says, not divine knowledge—as "some device of persuasion which will make one appear to those who do not know to know better than those who know" (Plato 1975, 459c). In the *Republic,* Plato draws a continuum of validity with a philosopher's divine knowledge (accurate, logical assessments of truth) at one extreme, ignorance (wholly inaccurate assessments of truth) at the other, and opinion (occasionally accurate but illogically derived assessments of truth) somewhere in between (Plato 1989c, 477a–b). Rhetoricians, Socrates says in the *Gorgias,* compellingly and unethically present opinion as knowledge—and they speak so persuasively that they help opinion trump God's truth. In the *Phaedrus,* therefore, Socrates maintains that only philosophers can legitimately use the persuasive powers of rhetoric. They have found God's truth and can use rhetorical strategies to lead the unenlightened to knowledge: "A man must know the truth about all the particular things of which he speaks and writes" (Plato 1928, 277b).

In the *Republic,* this marriage of philosophy and rhetoric takes ultimate form in the philosopher king: "Until, said I, either philosophers become kings in our states or those whom we now call our kings and rulers take to the pursuit of philosophy seriously and adequately, and there is a conjunction of these two things, political power and philosophical intelligence . . . there can be no cessation of troubles, dear Glaucon, for our states, nor, I fancy, for the human race either." Plato's philosopher king rules with absolute power, knowing that his version of reality is correct. And to bring an unruly populace to knowledge, or at least to an inability to debate or protest, Plato's Socrates gives that king exclusive license to lie and cheat, even within state-arranged breeding policies:

> The rulers then of the city may, if anybody, fitly lie on account of enemies or citizens for the benefit of the state; no others may have anything to do with it. . . . If then the ruler catches anybody else in the city lying . . . he will chastise him for introducing a practice as subversive and destructive of a state as it is of a ship. . . .

It seems likely that our rulers will have to make considerable use of falsehood and deception for the benefit of their subjects. . . . It follows from our former admissions, I said, that the best men must cohabit with the best women in as many cases as possible and the worst men with the worst in the fewest. . . . Certain ingenious lots, then, I suppose must be devised so that the inferior man at each conjugation may blame chance and not the rulers. (Plato 1989c, 473c–e, 389b–c, d, 459c–d, 460a)

Plato thus maintains that the soul of ignorance—or at least the soul of opinion—opposes knowledge, endangers the stable state, and must be crushed by fair means or foul. Filled with divine knowledge that must not be contested, the philosopher king must impose one view—his view—on his subjects.

Classical scholars haven't overlooked this breathtaking arrogance. George Kennedy (1994) writes that Plato is "one of the most dangerous writers in human history, responsible for much of the dogmatism, intolerance, and ideological oppression that has characterized western history" (41). Charles Kauffman (1994) labels Platonic rhetoric "totalitarian and repressive" (101). Edwin Black (1994) maintains that such rhetoric is a form of "social control" (98). Werner Jaeger (1944) calls it "uncompromising" (70), and George Yoos (1984) concludes that it "cannot function in negotiations and compromise" (85). Popper (1966) refuses to believe Socrates capable of such baseness, preferring to blame Plato's portrayal of the man: "It is hard, I think, to conceive a greater contrast than that between the Socratic and the Platonic ideal of a philosopher. . . . Plato libels his great teacher" (132, 150).

But what if Socrates did teach this repressive blend of philosophy and rhetoric to youth, if, figuratively, he did worship incontestable, divine ideas at the expense of social debate? In that case, those long-ago charges in Athens may well have been true in a sense unintended by the original accusers: Socrates did corrupt youth and worship false gods.

ARISTOTLE AND THE SOUL OF IGNORANCE

The problem with rhetoric as popularly conceived, Plato wrote, was that it dealt with opinion, with probability, rather than knowledge. He decried rhetoricians "who saw that probabilities are more to be esteemed than truths" (Plato 1928, 267a [*Phaedrus*]). However, his greatest student, the one he called "the reader"—Aristotle—eventually came to disagree with his teacher. In his *Rhetoric,* a subject he first taught while still studying in Plato's Academy, Aristotle writes that in some situations "exact certainty is impossible and opinions are divided." Rhetorical proofs, then, can be built on "probabilities," and decisionmakers must judge accordingly: "If you have no witnesses on your side, you will argue that the judges must decide from what is probable. . . . For they ought to decide by considering not merely what must be true but also what is likely to be true; this is, indeed, the meaning of 'giving a verdict in accordance with one's honest opinion'" (Aristotle 1954, 1356a, 1359a, 1376a, 1402b). Not surprisingly, Aristotle had rejected Plato's philosophy of divine ideas, those gold standards of certainty. In the *Metaphysics,* he concludes that such ideas are "tantamount with mere poetic metaphors" (Aristotle 1991, 1079b).

Rhetoric as an art, a subject of study, may well have begun in the mire of ignorance, opinion, and probability. Cicero tells us that Aristotle, in a work now lost, traced the beginning of formal rhetoric to Corax and Tisias of Syracuse, who invented a method of putting forth logical probabilities to help settle land disputes (Cicero 1970a, 12 [*Brutus*]). When Thrasybulus, a tyrannical ruler, had been overthrown in favor of a rudimentary democracy, citizens who had lost land to the tyrant argued, in the absence of property records, for the restoration of their property. As claims conflicted and certainty seemed impossible, Corax and Tisias began to establish how citizens could present and debate alternate probabilities. Though some scholars argue that Corax and Tisias were the same individual, there survives a perhaps-apocryphal anecdote of the master Corax suing the student Tisias for failure to pay for his instruction. Corax argued that, if his case prevailed, he won—and that, if Tisias' case prevailed, he still should win, for in winning Tisias had proved himself to be a learned rhetorician who should pay for his knowledge. Tisias allegedly countered that, if he won, he need not pay—and, if he lost, he proved that he hadn't learned enough of the art of probable arguments to merit payment.

With divine certainty impossible, Aristotle maintains that we have no choice but, in Plato's disdainful words, to "wander amid the multiplicities of multifarious things" (Plato 1989c, 484b [*Republic*]). Being Aristotle, he surveyed the history and practice of rhetoric and wrote what remains the most famous treatise on the subject, the three-book work commonly known as the *Rhetoric.* But, having tried to liberate ignorance, probability, and opinion from their banishment under Plato, Aristotle presents another problem in the battle for the soul of ignorance. Rather than advocating an art of probabilities that seeks the best option for all concerned, he presents a rhetoric bent on winning debates, one that turns an audience's ignorance against it. Rhetoric, says Aristotle in his famous definition, "is the faculty of observing in *any* given case the available means of persuasion" (1954, 1355b [emphasis added]). Thus, one of the debaters in Cicero's *De oratore* wonders whether Greek orators truly are able "in Aristotelian fashion to speak on both sides about every subject and by means of knowing Aristotle's rules to reel off two speeches on opposite sides of every case" (1970b, 3.21). In his essay "Mighty Is the Truth and It Shall Prevail?" Robert Wardy (1996) details Aristotle's "catalogue of fishy persuasive techniques" and concludes that, in the *Rhetoric,* Aristotle "subordinates truth to victory." At its worst, Wardy says, Aristotle's *Rhetoric* is "a rampant instance of Plato's worst nightmare" (74, 81, 79)—of rhetoric in the service not of divine truth but of falsehood.

The *Rhetoric* abounds with examples of Aristotle's slaps at ignorant audiences. In his essay "Aristotle's Knowledge of Athenian Oratory," J. C. Trevett (1996) catalogs some of the assessments of the Athenian hoi polloi: "In a number of passages he also belittles the capabilities of the audience before whom such speeches are delivered. For example, the judge should be assumed to be a simple person who is incapable of following a complex argument . . . ; maxims are useful because of the uncultivated mind of the audience . . . ; the uneducated are more persuasive than the educated before a crowd . . . ; remarks to the audience in the proem of a judicial speech are outside the real argument and are addressed to hearers who are morally weak" (378). This from a man who still holds sway in college ethics courses as the champion of moderation, who advocates seeking and practicing virtuous behavior to become, eventually, inherently virtuous? Modern scholars see no contradiction, maintaining that, in the *Rhetoric,* Aristotle analyzed what was, not what should be; he offered a descriptive, rather than a prescriptive, rhetoric. Kennedy (1994) compares the *Rhetoric* to Aristotle's "dispassionate" analyses of plants and animals (56). In *Rhetoric in Greco-Roman Education,* Donald Clark (1957) holds that Aristotle "tried to see the thing as in itself it really was . . . [and] endeavored to devise a theory of rhetoric without moral praise or blame" (24).

For the Aristotelian rhetorician, ignorance is not finite and curable; philosopher kings armed with God's ideas cannot eradicate it. But this realization, heartening if only in its rejection of Platonic repression, doesn't imbue ignorance with a soul worthy of a worldview. This realization renders ignorance a shameful tactic, a miserable shortcoming in a dull audience—people whom Mencken called the booboisie—that allows an unscrupulous rhetorician to pander, deceive, and win.

ISOCRATES AND THE SOUL OF IGNORANCE

Isocrates lived ninety-eight years. He was the contemporary of Socrates, Plato, and Aristotle, born thirty-seven years before Socrates' death, nine years before the birth of Plato, and fifty-two years before Aristotle. He may have been a student of Socrates'. His school, famed throughout Greece, competed with those of Plato and Aristotle.

Though less famous today than his contemporaries, Isocrates went toe-to-toe with the rhetorics of Plato and Aristotle—and he won. Says Cicero: "Then behold Isocrates arose, from whose school, as from the Trojan horse, none but real heroes proceeded" (1970b, 2.22 [*De oratore*]). Henri Marrou (1956), a Platonist and the author of *A History of Education in Antiquity*, virtually gnashes his teeth in declaring, sorrowfully but gamely: "On the whole, it was Isocrates, not Plato, who educated fourth-century Greece and subsequently the Hellenistic and Roman worlds" (79). Kenneth Freeman (1907), in *Schools of Hellas,* concludes: "The pupils of Isokrates became the most eminent politicians and eminent prose-writers of the time" (186).

Aristotle allegedly began his own classes in rhetoric with the declaration that it was shameful to remain silent in the face of Isocrates' teachings. Plato, through his Socrates, may have tried to win over his rival with a cryptic remark—sincere and/or sarcastic?—at the end of the *Phaedrus:*

> Isocrates is still young, Phaedrus, but I don't mind telling you the future I prophesy for him. . . .
>
> It would not surprise me if with advancing years he made all his literary predecessors look like very small-fry—that is, supposing him to persist in the actual type of writing in which he engages at present—still more so, if he should become dissatisfied with such work, and a sublimer impulse lead him to

do greater things. For that mind of his, Phaedrus, contains an innate tincture of philosophy. (1989b, 278e–279a)

Those "greater things" surely meant Platonic things, which Isocrates fiercely rejected to the end. Isocrates battled (and probably lost to) Plato on the meaning of philosophy. To Demonicus, a young friend, Isocrates wrote: "You are ripe for philosophy and I direct students of philosophy" (1991a, 3). For Isocrates, philosophy and rhetoric were identical, and he preferred the term *philosophy*. But, unlike Plato, Isocrates founded his philosophy on uncertainty, opinion, and probability—and, unlike the amoral rhetoric described by Aristotle, Isocratean philosophical rhetoric aimed at social good, not personal victory. Isocrates refused to use an audience's ignorance against it.

Isocrates is blunt in his rejection of Platonic certainty:

It is appropriate for me . . . since I hold that what some people call philosophy is not entitled to that name, to define and explain to you what philosophy, properly conceived, really is. My view of this question is, as it happens, very simple. For since it is not in the nature of man to attain a science by the possession of which we can know positively what we should do or what we should say, in the next resort I hold that man to be wise who is able by his powers of conjecture to arrive generally at the best course, and I hold that man to be a philosopher who occupies himself with the studies from which he will most quickly gain that insight. (1992a, 270–71 [*Antidosis*])

Having squared off against Plato, Isocrates next disparages the coldly analytic assessments, collected by Aristotle in the *Rhetoric,* of tactics that will win debates. In their place, Isocrates offers a nobler purpose for rhetoric and rhetoricians, the concept of true "advantage," of serving society by finding the best decision or best course of action:

I do hold that people can become better and worthier if they conceive an ambition to speak well, if they become possessed of the desire to be able to persuade their hearers, and, finally, if they set their hearts on seizing their advantage—I do not mean "advantage" in the sense given to that word by the empty-minded, but advantage in the true meaning of that term. (1992a, 275–76 [*Antidosis*])

Nothing in the world can contribute so powerfully to material gain, to good repute, to right action, in a word, to happiness, as virtue and the qualities of virtue. For it is by the good qualities which we have in our souls that we acquire also the other advantages of which we stand in need. . . .

But I marvel if anyone thinks that those who practise piety and justice remain constant and steadfast in these virtues because they expect to be worse off than the wicked and not because they consider that both among gods and among men they will have the advantage over others. I, for my part, am persuaded that they and they alone gain advantage in the true sense. (1992b, 32, 33–34 [*On the Peace*])

In seeking advantage, the Isocratean philosopher/rhetorician challenges his own beliefs by considering the interests of others. To the children of Jason, a former ruler of Pherae, Isocrates wrote: "I should be ashamed if I should be thought by any either to be neglectful of you on account of my city, or on your account to be indifferent to the interests of Athens" (1986, 14 [*To the Children of Jason*]). In *To Nicocles,* he tells the new ruler of Cyprus: "Regard as your most faithful friends, not those who praise everything you say or do, but those who criticize your mistakes" (1991b, 28). In *Speaking for the Polis,* Takis Poulakos (1997) says of Isocrates: "He changed the art of rhetoric from the way it was when he inherited it. The version of rhetoric he left behind is, unequivocally, a rhetoric for the polis [community]. . . . He mobilized his version of rhetoric in order to advocate courses of action that safeguarded the interests of citizens and promoted the general welfare of the city-state" (3–4).

Isocrates' vision of philosophy did not prevail. But his version of rhetoric shaped the leaders of a generation and influenced the great Roman rhetoricians Cicero and Quintilian, both of whom stressed the moral character and civic duties of the rhetorician. Marrou (1956) maintains that Isocrates aimed rhetoric at philosophy but hit a different, perhaps nobler target: "In the hands of Isocrates rhetoric is gradually transformed into ethics" (89). In establishing a deferential rhetoric that distrusts certainty, solicits dissent, and sees advantage in working for the common good, Isocrates redefines the soul of ignorance in classical Athens. He defines an ignorance worthy of being a worldview.

TOWARD A MODERN IGNORANCE-BASED WORLDVIEW

So what happened? Isocrates left us a legacy, in his words, "more noble than statues of bronze" (1992a, 7 [*Antidosis*]). How did we become so Platonic, so heedless of dissent, and so certain that we can alter our environment and master the consequences? In the sixteenth century, the French scholar Peter Ramus ripped logic from rhetoric, leaving the latter only delivery and figures of speech—a revision that prevailed throughout Europe and certainly undermined the injection of critical thinking into public debate. Or perhaps the Enlightenment, the industrial revolution, and the advances of science dragged us back to a modified Platonism, to a belief that we could attain incontestable certainty as a prelude to action. We should study and learn how we fell away from the Isocratean ideal.

But here we are again, doubting certainty. Challenging Plato. Suggesting that deferential toe dipping may serve us better than arrogant, damn-the-torpedoes confidence. In his *Pensées,* the French philosopher Blaise Pascal (1910) writes: "The sciences have two extremes, which meet. The first is the pure natural ignorance in which all men find themselves at birth. The other extreme is that reached by great intellects, who, having run through all that men can know, find they know nothing and come back again to that same ignorance from which they set out; but this is a learned ignorance which is conscious of itself" (114–15). After two-plus millennia of philosophical exploring, we have come back again to the debates of classical Athens. As we "wander amid the multiplicities of multifarious things," can we relaunch the battle for the soul of ignorance? For the first time, do we, in learned ignorance, know where we are and what's at stake?

If so, let's not blow it again.

REFERENCES

Aristotle. 1954. *The Rhetoric and the Poetics of Aristotle.* Translated by W. R. Roberts and I. Bywater. New York: Modern Library.
———. 1991. *Metaphysics.* Translated by J. H. McMahon. Amherst, NY: Prometheus.

Black, E. 1994. "Plato's View of Rhetoric." In *Landmark Essays on Classical Greek Rhetoric,* ed. E. Schiappa, 83–99. Davis, CA: Hermagoras. Reprinted from *Quarterly Journal of Speech* 44 (1958): 361–74.

Cicero. 1970a. *Brutus.* In *On Oratory and Orators,* ed. and trans. J. S. Watson, 262–367. Carbondale: Southern Illinois University Press. Translation originally published in 1878.

———. 1970b. *De oratore.* In *On Oratory and Orators,* ed. and trans. J. S. Watson, 1–261. Carbondale: Southern Illinois University Press. Translation originally published in 1878.

Clark, D. L. 1957. *Rhetoric in Greco-Roman Education.* New York: Columbia University Press.

Freeman, K. J. 1907. *Schools of Hellas: An Essay on the Practice and Theory of Ancient Greek Education.* London: Macmillan.

Isocrates. 1986. *To the Children of Jason.* Translated by LaRue Van Hook. In *Isocrates,* 3:433–43. Cambridge, MA: Harvard University Press. Translation originally published in 1945.

———. 1991a. *To Demonicus.* Translated by G. Norlin. In *Isocrates,* 1:1–35. Cambridge, MA: Harvard University Press. Translation originally published in 1928.

———. 1991b. *To Nicocles.* Translated by G. Norlin. In *Isocrates,* 1:37–71. Cambridge, MA: Harvard University Press. Translation originally published in 1928.

———. 1992a. *Antidosis.* Translated by G. Norlin. In *Isocrates,* 2:179–365. Cambridge, MA: Harvard University Press. Translation originally published in 1929.

———. 1992b. *On the Peace.* Translated by G. Norlin. In *Isocrates,* 2:1–91. Cambridge, MA: Harvard University Press. Translation originally published in 1929.

Jaeger, W. 1944. *The Conflict of Cultural Ideals in the Age of Plato.* Vol. 3 of *Paideia: The Ideals of Greek Culture.* Translated by G. Highet. New York: Oxford University Press.

Kauffman, C. 1994. "The Axiological Foundations of Plato's Theory of Rhetoric." In *Landmark Essays on Classical Greek Rhetoric,* ed. E. Schiappa, 101–16. Davis, CA: Hermagoras. Reprinted from *Central States Speech Journal* 33 (1982): 353–66.

Kennedy, G. A. 1994. *A New History of Classical Rhetoric.* Princeton, NJ: Princeton University Press.

Marrou, H. I. 1956. *A History of Education in Antiquity.* Translated by G. Lamb. New York: Sheed & Ward.

Pascal, B. 1910. *Thoughts.* Translated by W. F. Trotter. In *Blaise Pascal: Thoughts and Minor Works* (Harvard Classics, vol. 48), ed. C. W. Eliot, 7–322. New York: P. F. Collier & Son.

Plato. 1928. *Phaedrus.* Translated by H. N. Fowler. In *Euthyphro, Apology, Crito, Phaedo, Phaedrus,* 405–579. Cambridge, MA: Harvard University Press. Translation originally published in 1914.

———. 1975. *Gorgias.* Translated by W. R. M. Lamb. In *Lysis, Symposium, Gorgias,* 247–533. Cambridge, MA: Harvard University Press. Translation originally published in 1925.

———. 1989a. *Apology.* Translated by H. Tredennick. In *Plato: Collected Dialogues,* ed. E. Hamilton and H. Cairns, 3–36. Princeton, NJ: Princeton University Press.

———. 1989b. *Phaedrus.* Translated by R. Hackforth. In *Plato: Collected Dialogues,* ed. E. Hamilton and H. Cairns, 475–525. Princeton, NJ: Princeton University Press.

———. 1989c. *Republic.* Translated by P. Shorey. In *Plato: Collected Dialogues,* ed. E. Hamilton and H. Cairns, 575–844. Princeton, NJ: Princeton University Press.

Popper, K. 1966. *The Open Society and Its Enemies.* 5th ed. London: Routledge.

Poulakos, T. 1997. *Speaking for the Polis: Isocrates' Rhetorical Education.* Columbia: University of South Carolina Press.

Quintilian. 1980. *The Institutio oratia of Quintilian.* Translated by H. E. Butler. Cambridge, MA: Harvard University Press. Translation originally published in 1920.

Russell, B. 1945. *A History of Western Philosophy.* New York: Simon & Schuster.

Trevett, J. C. 1996. "Aristotle's Knowledge of Athenian Oratory." *Classical Quarterly* 46:371–80.

Wardy, R. 1996. "Mighty Is the Truth and It Shall Prevail?" In *Essays on Aristotle's "Rhetoric,"* ed. A. O. Rorty, 56–87. Berkeley and Los Angeles: University of California Press.

Yoos, G. 1984. "Rational Appeal and the Ethics of Advocacy." In *Essays on Classical Rhetoric and Modern Discourse,* ed. R. J. Connors, L. Ede, and A. A. Lunsford, 82–97. Carbondale: Southern Illinois University Press.

Yunis, H. 2003. *Written Texts and the Rise of Literate Culture in Ancient Greece.* Cambridge: Cambridge University Press.

Choosing Ignorance within a Learning Universe

Peter G. Brown

The Universe is wider than our views of it.
                                        —Henry David Thoreau

I had the good fortune in recent years to canoe down the Old Factory River in central Quebec with people who were experienced on and with the river and also steeped in the East Cree culture of the region. As I awoke each morning, I asked a number of questions so that I could think about and plan for the day. Here is how it went:

Me: How many portages will there be?
Response: Don't know yet.
Me: Are the rapids big?
Response: Will have to see.
Me: Is it likely to rain?
Response: Sometimes does.

My frustration grew as I realized that I knew little or nothing more than when I started. I had trouble imagining how anyone could possibly be so uninformed about what was around them. It took me years to understand that it was actually me who was uninformed. I wanted to know things that could not be known yet. How many portages there would be depended on the speed and skill of the paddlers and the number of interesting diversions and mishaps that occurred along the way. The size of the rapids would depend on which rapids we encountered, and that would depend on which portages we made. I did not understand the limitations of knowledge. I did not understand the wisdom of choosing a kind of ignorance based on respect.

OVERVIEW

In this essay, I examine the fusion of scientific knowledge with an ethic of power. I argue that this conception leads, ironically, to increases in ignorance. It is, therefore, dialectical—bringing forth its own opposite. I then turn to an alternative prescriptive concept of the relation between ethics and knowledge derived from Albert Schweitzer's concept of reverence for life. Here, we can ground a kind of ignorance that leads to true power and profound freedom.

Knowledge as Power

There are many kinds of knowledge: theoretical knowledge, traditional knowledge, knowledge of our bodies ("I have a toothache"), and sexual knowledge, to name just a few. Here, I concentrate on the consequences of the merger of scientific knowledge, its resultant technologies, and capitalism with an ethic of the quest for power over nature and, as it typically turns out, over other allegedly "lesser" peoples. This unholy alliance is now in full swing around the globe, extinguishing species at an accelerating rate, while further undercutting many ways of life more adapted to survival and flourishing than the presently hegemonic Western culture.

It is important to make the main points of my thesis clear. My critique is not of science itself but of the belief systems in which it is now largely *entwined*. I do not argue for a science free of ethics, metaphysics, and theology. This is neither possible nor desirable. Of science, as of other human endeavors, it is always possible to ask, What is it for? The relation between science and ethics, metaphysics, and theology is reflexive: they mutually influence and shape each other. Science helps us understand who we are and where we came from; this understanding in turn shapes but does not wholly determine our inclinations, passions, and ideas about where we might and should go, our ideas about the nature of reality, and our proximate and ultimate commitments.

Knowledge-Power and Ignorance

In our pursuit of knowledge for power, we increase ignorance in six ways. I do not know whether there is a net increase in ignorance. That would be very difficult to figure out. My argument is that the *kinds* of ignorance generated account in part for humanity's current and growing dysfunctional relationship with life and the world. First, this kind of knowledge is inherently unreflective about its own ethical, metaphysical, and theological assumptions. Second, it is inherently limited, even self-limiting, and is shifting from a generator of benefits to a distributor of risks. Third, as science has undercut its own legitimating narratives, it has rendered it more and more difficult to say what science in particular, and society in general, should do and which technologies should be developed and which neglected. Fourth, the abstractions that make up scientific knowledge inadequately foresee the consequences of its applications in complex adaptive systems. Fifth, the privileged knowledge of Western science often displaces the much more functional, but less theoretically complex, understandings of nature and culture of traditional peoples. Finally, the current manifestation of the knowledge-as-power project impedes the very way the universe itself learns: evolution. This last and key aspect of the sixfold knowledge problematic—the loss of evolutionary learning—is last but by no means least. It represents a catastrophe for life on earth if it stays on its present course. It also offers an opportunity. For, without an evolutionary outlook, seeing ourselves as embedded in a vast yet progressive unfolding, we become doomed to remain trapped in the present tragic spectacle of our own making, lonely exiles in a diminished world.

Knowledge in Service to Life

Our primary orientation to the world should be ethical *and mystical* (characterized by a sense of awe and joy)—at once grounded in emotion and thought. Scientific knowledge, its related technologies, and economics should be in service to what Albert Schweitzer called an ethic of reverence for life. I do not reject the importance of understanding ourselves and the world around us, but I do take fundamental issue with the domination project and with many of the unscientific assumptions on which it rests. I agree with Schweitzer that an exit from these ills of our time lies outside the mainstream of the Western tradition. Neither the mainstream Judeo-Christian worldview nor its seventeenth-century successor, the mechanical-material worldview of the Enlightenment, serves us well in reaching a right relationship with life and the world. Biblically based claims to be the chosen species to which the earth should be subjugated have proved to be artifacts in service of global ecological vandalism, though, regrettably, it is not the only cultural heritage with such a tragic legacy. Similarly, the alleged superiority of Western "enlightened" man and the attendant idea of the "white man's burden" are also conceits to legitimate the subjugation of others. Our minds are not the clear light of reason, as the Enlightenment suggests, but are dependent on, emergent from, and entangled with our bodies and our passions—as are the minds of other animals. Schweitzer's ethic of reverence for life, and the life as a physician that he lived by it, offers an exit from the present path toward the global extinction calamity and the radical pauperization of hundreds of millions of persons. It offers an ethic to ground a responsible epistemology, an epistemology that recognizes its own ignorance.

KNOWLEDGE AS POWER

In the seventeenth century, the progress project assumed center stage with the promise that humanity's lot could be improved through the systematic and intentional advancement of science, technology, and the domination of nature. We could reduce disease, address malnutrition, and lighten the burden of physical labor. Its father was Francis Bacon, who saw science as humanity's means to "recover that right over nature which belongs to it by divine bequest."[1] But the consequences were not just the sciences but, as Carolyn Merchant has argued in *Reinventing Eden,* technology and capitalism as well. The progress project laid the foundations some four centuries ago for the current, largely unrecognized crisis in knowledge and ethics. The legitimating *nonscientific* narrative for Bacon came from the Judeo-Christian tradition. Bacon proposed that through knowledge we can control nature, making "her" subservient to our purposes by making a global Garden of Eden. We can, thus, restore ourselves to the place God gave us before the Fall from paradise depicted in the first chapters of Genesis. Progress is, thus, understood as the relentless movement to restore our rightful place as the masters of creation. In the words of an international agricultural giant, we should aim at "making the world one big farm field."[2] Robert Faulkner describes how Bacon placed the agenda of science for power directly in the Judeo-Christian narrative:

> The *New Atlantis* is Bacon's poetic image of a new world of increasing satisfaction, and it transforms, while appropriating, the

future-oriented promise of Christ. It reminds us of a promised land and messiah by the Hebrew connotations of "Bensalem" and Joabin, by various identifications of Bensalem with Savior, Lord, and Virgin, by the Europeans' identification of their welcome with salvation. . . . Yet the new land promises not milk and honey in heaven, but Gatorade and Nutra-Sweet on earth; not mortality, innocence, and union with God, but security, luxury, and power. The provider of future bliss is not providence but the state, a progressive science, and their founders.[3]

Francis Bacon is often criticized for having a naive conception of scientific method, one that relies too much on induction and too little on deductive inference. He is reviled by ecofeminists as legitimating "raping" a feminized earth. What is beyond dispute is his role as one of the founders of the modern progress project. He set his foot firmly against the scholastic tradition of trying to understand the world through metaphysical and theological interpretation of ancient and more contemporary texts. He deplored the practice of deduction without careful attention to empirical premises.

Taking Back the Garden

Bacon cast his project firmly within the metaphor of the Fall of man from the Book of Genesis. By eating of the fruits of the tree of knowledge, we humans bring about not our own fall and expulsion from the garden but that of nature as well. We thus live in a state of sin and in a world degraded by our own acts. Firmly in the tradition of "some hair from the dog that bit you," Bacon proposes that *mankind* can restore its place in the garden by the systematic accumulation of scientific and technical knowledge, but he assumes—incorrectly as we will see—that this project rests on a solid ethical foundation of human ownership of the earth.

Now, some four hundred years into the project, how should we assess it? Typically, one cites reduction in disease, longer human lives, expanded food supplies, less painful dentistry, more rapid communication and travel, and the like. Yet, for a number of reasons, we should be less confident than we are. First, a crucial part of the legitimating mainstream narrative is to implicitly or explicitly paint the societies outside it as backward: having lives that are, to use Hobbes's phrase, "nasty, brutish and short." Not mentioned in these set-piece and often empirically naive arguments are the successful societies that have been annihilated and the good and satisfying lives lived in those few that remain—now increasingly well documented in the anthropological literature.[4] Second, also omitted is the role of abundant refined sugar in expanding the need for dentistry and the epidemic of obesity and diabetes brought on by allegedly improved food supplies. Nor is there candid recognition of the coming epidemics of cancer, heart disease, and emphysema sure to follow increased smoking and the surging increase in urban air pollution in many parts of the world. Third, there are also objections in principle—from detractors like Thoreau and, more recently, the ecofeminists. Thoreau wryly noted that improved communication between New England and England itself did nothing to ensure that anyone had anything important to say. For him, two steps forward almost always meant at least one step backward. The trivial hubbub of the nineteenth century blocked humankind's experience and knowledge of the transcendent. Ecofeminists have mounted sustained and telling critiques of the progress project—a subject to which I will return when I discuss the philosophy of history. But, despite these well-taken critiques, in the main it has been the Baconian agenda, a powerful fusion of metaphysics, moral and theological metaphors, and enormous investments in science and technology in service to market forces—a triumvirate that assigns a low priority to the health problems of the world's poor because of their low purchasing power[5]—that has been triumphant in our time.

HOW KNOWLEDGE INCREASES IGNORANCE

Science's Hidden Metaphysics/Theology

Science makes certain substantive assumptions to which, owing to its emphasis on the empirical and "objective," it is slow to notice and correct. It is unavoidably embedded in metaphysical and theological matrices that shape its questions and, thus, its findings as well as *its assumptions about itself.* Two of these are *us/it* and *internal/external.* The first of these—that humans are apart from nature, from everything else physical—is deeply embedded in the dualism of Judeo-Christian thought, which sees mankind as created in the image of God and as different *in kind* from all other animals and from nature itself. Another draught of mischief is introduced in the Enlightenment, for example, by

Descartes' insistence that it is only humans who have minds. The activity inside the minds of human beings is placed *apart from the rest of the cosmos.*[6] This assumption slowed empirical investigation of the mental lives of other species.[7] Why investigate something—the minds of other species—that is not there? This lacuna has served, in part, to legitimate the rise of animal "science" departments that treat nonhuman animals as mere objects of human utility and to legitimate the "captured lives" of hundred of millions of chickens, pigs, and cattle.

The idea that the human mind is somehow something apart from nature also slowed the development of cognitive science concerning the nature of human consciousness. As George Lakoff and Mark Johnson point out: "What is subject to physical law can be studied scientifically—the physical world, including biology. But, being radically free and not subject to the laws of physical causation, the mind is not seen as amenable to scientific study. A different, 'interpretative' methodology is supposedly required for the human sciences. For this reason, cognitive science has not been taken seriously within traditional humanistic fields of study."[8] As a result, at the turn of the millennium we are well over two thousand years late in achieving one of the fundamental goals of the Western tradition, Socrates' admonition: "Know thyself." An empirically based understanding of our own consciousness and how it fits into what we know about the nature of matter from quantum physics to evolutionary biology is now just beginning.

Unjustified methodological assumptions also abound and shape our understanding of what we can know, as exemplified in the idea of *inside* and *outside.* It is widely assumed by scientists that scientific knowledge is objective and that this certifies its authenticity. Moral sentiments and ethical statements are taken, by contrast, to be *subjective* and, therefore, unscientific, an assumption about the nature of mental life and moral judgments typically accepted without argument or evidence. Most, but not all, scientists are unwilling to engage in a debate about the ethics of their own work. Those who do typically enjoy a status that allows them—E. O. Wilson and Aldo Leopold, for example—to break with the canon that science and ethics should never be mixed. The tragic result of the refusal of mainstream science to be concerned with ethics is the pell-mell development of science and technology with little or no appraisal of its morality, without noticing that the whole enterprise rests on beliefs and metaphors inconsistent with modern science itself. This ethically and metaphysically ungrounded scientific and technical process is now destabilizing the very things that make life possible, such as the earth's ozone layer. Is the morally ignorant scientific and technological project reaching some kind of finale?

The Red Queen

The scientific enterprise also systematically increases ignorance in both its synthetic and its analytic activities. I will use chemical engineering to illustrate these. In the pursuit of the product novelty needed to stimulate increased consumption, chemical engineers invent new compounds all the time. This activity is at once analytic and synthetic. Existing compounds are broken down into their molecular parts (the analytic dimension is established) and then recombined to make new substances such as flame retardants (the synthetic phase).

Regulatory bodies such as the Environmental Protection Agency in the United States and Environment Canada track several thousands of these compounds primarily for their effects on human health. Let's say that there are two thousand of them (there are many more, but this arbitrary number is more than sufficient to make the point). It is now commonplace to note that we need to know about their effects both individually and in concert with each other. Two thousand times two thousand is 4 million. And then it is noted that we need to know their effects on a number of systems: for example, circulatory, nervous, reproductive, lymphatic, etc. Let's suppose that there are four of these. So we now have 16 million classes of effects to look after. *Add* their effect on nonhuman organisms and ecosystems. We can see that we have created a problem not unlike that experienced by the Red Queen in *Alice in Wonderland.* We have to keep running faster and faster to stay in the same place—or, in this case, to try to keep up our "knowledge" with the receding horizon of unforeseen consequences from the progress project. We don't have to determine whether there is a net increase in ignorance to see that the expansion and application of knowledge leads to an ignorance of what we need to know to protect ourselves and other living beings that grows much faster than knowledge itself does.[9]

The conventional image we have is that knowledge is expanding and ignorance receding. Eventually, knowledge will expand to even the dark corners of the universe, and all mystery and all need for religious faith will be eliminated. This conception is strongly misleading. Indeed, we are faced with an combinatorial explosion: knowledge increases arithmetically, while ignorance increases exponentially. Before these chemical compounds were created, the particular interactions now possible could not have happened; hence, a new opportunity for not knowing—ignorance—has been produced. This proliferation of compounds might be called molecular imperialism. Green chemistry, industrial ecology, and biomimicry are attempts to slow down the race of the Red Queen. They attempt to reduce the number of compounds that are orthogonal to nature, or at least to prevent their release. In a way, these movements are an attempt to reground knowledge from a point of view of working with natural systems as opposed to seeking to have power over them.

How Science Helped Create the Moral and Metaphysical Orphanage of Our Time

Paradoxically, as the theological and metaphysical underpinnings of the Baconian project of a return to the garden have decayed in large part as a result of several centuries of scientific findings, the project has continued to gain momentum and now shapes even the far corners of the earth and the distant future. More paradoxical still is the scientific progress project's role in destabilizing the enlightenment project that reifies rationally based knowledge as the defining dimension of the person. Recent developments in cognitive science paint a very different image of the person than that found in the eighteenth-century philosophers who formulated the Enlightenment worldview. As Lakoff and Johnson note: "The traditional Western view of the person is . . . at odds *on every point* with the fundamental results from neuroscience and cognitive science. . . . An actual human being has neither a separation of mind and body, nor Universal Reason, nor an exclusively literal conceptual system, nor a monolithic consistent worldview, nor radical freedom."[10]

As a result, the progress project is now unleashed in the metaphysical orphanage of our time, having slain its own parentage from the Book of Genesis and the Enlightenment. And, owing to the dictum that moral judgments have no place in science, little or nothing has been done to replace these frameworks with alternatives that could be used in assessing what to do. Technological innovations such as in the field of biotechnology proceed without reference to agreed-on standards by which to judge their desirability. This triggers what Ulrich Beck has called a reflexive response, where the purposes of science itself are subject to dispute.[11] Indeed, without moral standards, it is even impossible to say what is a benefit and what is not. Is the growing of headless "humans" as organ banks desirable or not? We have no means to judge. *Hence, science creates the crisis of its own legitimacy.*

But the situation is worse, far worse. It is not just science that is rudderless. It is our so-called civilization itself. Our moral standards for judging the future directions of the whole human enterprise are metaphysical orphans. Our Western moral systems are ascientific—they were constructed before, and, hence, necessarily without reference to, the findings of the last five hundred years of science. In a way, they are maps, but, like those possessed by the characters in Peter Sellers's film *The Magic Christian* (1969), they are not maps of where we are. We are desperately in need of moral systems grounded in and consistent with what we know about who we are and where we came from.

It is not just our standards that are lacking. Partly as a result of the success of the progress project, we are often ignorant of the very information necessary to live an ethical life. What went into the meal I just consumed, who was affected and how, where did the financing come from, what were the ecological consequences of growing, processing, transporting, and packaging it? Getting answers to questions like these is impractical at best. The short feedback loops between production and consumption of traditional societies, never perfect in themselves, are lengthened. Hence, the complexities of globalization render problematic, but not necessarily impossible, one of the most essential features of ethical life: What are the consequences of what I do? This failure is a major cause of the moral and spiritual crisis afflicting Western civilization correctly diagnosed by Schweitzer nearly a century ago.

Irreducible Ignorance: Surprise, Novelty, and Knowledge as Ignorance

If we had better knowledge, would we have less ignorance? In a way, yes. If we had knowledge grounded in complex systems and chaos theory, we would likely bring about fewer negative consequences. But, since theories and concepts are always simplifications of reality, they will never be as complex as reality itself. So there are ways in which our knowledge is always partial. There will be things that escape our net of abstractions. Tacit knowledge, the role of hunch, experience, savvy, and the like, reduces, but does not eliminate, this problem. We have to remain open to surprise: "'Chaotic' systems have an infinite sensitivity to the initial conditions imposed. The slightest deviation from the intended starting point generates a dynamical outcome which is entirely uncorrelated with the desired outcome."[12]

There is also the problem of novelty. In observing organisms in the process of evolution, phenotypes express an underlying genotype. But the genotype is modified through time—one of the primary mechanisms of evolution. Hence, there are new sets of initial conditions coming into being all the time. We are unable to predict the direction of these novel events.

There is also ignorance built into knowledge itself. This is so in three ways. First, much of what we know is expressed in language as opposed to mathematics. When Churchill noted that the English and the Americans were two people separated by a common language, he recognized that words could admit of very different meanings. Knowledge is always translated through our partially individualized worldviews, being irreducibly subject to distorted perception and the ambiguity of subjectivity. So much of what we know is notoriously ambiguous. Second, what we know is not that something is true but, rather, that it has not been falsified. So we don't know whether our beliefs are true—they always remain open to further testing. We never reach complete certainty. Third, "in 1931 Gödel proved that any axiomatic system which generates a formal basis for arithmetical operations must allow at least one theorem that, within the proof structure allowed by the axioms, can neither be proved nor disproved."[13] For example, in Euclidean geometry, the "common notion" that "the whole is greater than the part" is not subject to proof *within* the Euclidean system. There are always aspects of experience that we cannot penetrate with reason.

So, in some of these cases, knowledge will always be inadequate. This is an argument not for ignorance but for humility about our "knowledge." It should call into question our ability to master the world through the accumulation of knowledge and understanding.

How Knowledge Displaces Wisdom

The conception of knowledge as power also leads to ignorance in another way. The conceit that we of the industrialized West have a true understanding of the world leads to the view that our ways of knowing and doing are superior to those of "backward" traditional people.[14] Our knowledge—combined with technologies—carries, or is alleged to carry, more legitimacy than that of these peoples. As our ways displace them, ways of living far less damaging and far *more* respectful than ours are displaced. Many are irretrievably lost in the rush of modernity and markets. Fortunately, many traditional peoples are consciously working to reconstruct and strengthen their traditions, though often with very mixed results.

Killing How the Universe Learns

Another way the current knowledge-as-power regime increases ignorance is in reducing evolutionary variability. Darwin described certain aspects of the evolutionary processes on the earth, but twentieth-century science has shown that evolution is a property of the universe. The universe itself is an unfolding process of self-adaptation. This has been formulated by the Jesuit paleontologist Pierre Teilhard de Chardin in *The Phenomenon of Man*[15] and by chemists and physicists alike. For example, Eric Chaisson in *Epic of Evolution* describes the seven stages of cosmic evolution since the big bang as the particle epoch, the galactic epoch, the stellar epoch, the planetary epoch, the chemical epoch, the biological epoch, and the cultural epoch and refers to life on the earth as "a meaningful factor in the future evolution of the universe."[16] The sharp reduction in species numbers now under way, escalating as the juggernaut of progress and control transforms even the antipodes, and certain to rapidly accelerate as a result of climate change, reduces the pathways of future evolution—*the very way that the universe itself learns.*

BEGINNING WITH RESPECT AND AWE

We need to begin anew—with an ethic and a view of knowledge based on, but not wholly reducible to, the worldview provided by contemporary science. In the first half of the twentieth century, both Albert Schweitzer and Aldo Leopold took giant steps in this direction. Both devised ethics of deference and respect—not of mastery and control.

Schweitzer's quest for an ethical point of departure was both personal and public. He sought to give meaning and direction to his own life and to provide a means of rescuing the civilization that he saw in a state of collapse around him. He describes going slowly upstream on a steamer in Africa in 1915:

> Lost in thought I sat on the deck of the barge, struggling to find the elementary and universal conception of the ethical which I

had not discovered in any philosophy. Sheet after sheet I covered with disconnected sentences, merely to keep myself concentrated on the problem. Late on the third day, at the very moment when, at sunset, we were making our way through a herd of hippopotamuses, there flashed upon my mind, unforeseen and unsought, the phrase "Reverence for Life." The iron door had yielded: the path in the thicket had become visible. . . . Now I knew that the world-view of ethical world and life-affirmation, together with its ideals of civilization, is founded in thought.[17]

Reverence for life is not sentimentalism but a positive disposition toward other life-forms. Schweitzer was willing to make the choices to keep some life-forms alive while sacrificing others. It grew from a state of ethical mysticism that included recognition of unity with other *living* things. It was at once a withdrawal from the world and an affirmation of our life in it and our obligation to improve life's prospects. Like the basic music chord, which requires three notes played simultaneously to exist, ethics has three elements: we affirm ourselves and our own will to live, we affirm others and their will to live, and, at the same time, we hold back from the impulses in us to dominate and consume.

There is, of course, a great danger in this approach to life and ethics—the problems associated with ethical intuitionism. You can claim the primacy of your intuitions, and I can claim the primacy of mine, and there is no ground for deciding between the two. Moreover, ethical mysticism is more dangerous than intuitionism because it can claim a divine mandate and result in religious fanaticism. Then, in the name of improving the world, you end up bringing strife and discord. To avoid these pitfalls, other ways must be found to support our beliefs.

Schweitzer offered five reasons in support of reverence for life. First, he held that it is consistent with the two commandments of Jesus. Schweitzer took the commandment to love your neighbor to refer to all living things and the love of God to refer to all of creation. Second, reverence for life is enriched by and finds support in the great works of nineteenth- and early-twentieth-century philosophies of the will—particularly Nietzsche, Schopenhauer, Bergson, and Goethe. Unveiling the will to live in ourselves is an extension of the self in the direction of the creative will of the universe as, for example, in Schopenhauer's conception. Third, Schweitzer also found corroboration in the works of Adam

Smith and David Hume, who emphasized the role of sympathy in morality to share the feelings and the experience of the other. Fourth, he is drawing on the tradition of mysticism—a sense of the holy and our finiteness in its presence. It is also a call to action. Schweitzer's personal calling was to atone for the sins of the European colonialists in Africa. Fifth, he found support in the evolutionary biology of Charles Darwin, who argued that all differences between persons and other life-forms are matters of degree, not kind. Any attempt to establish a sharp moral boundary between humans and other species will fail.[18]

These five reasons can be thought of as a matrix of support—such as when a spaceship lands on the moon, carefully puts down its feet to level itself, and begins its work from a new perspective and with a new horizon. We can now add one more. In his *Philosophy of Civilization* (1923), Schweitzer tells us that a successful ethic for civilization must be grounded in what he calls a worldview—an interpretation of humanity's place in the universe. Yet he despairs of finding a way to connect an ethic of reverence for life to such a worldview. In the 1920s, he thus lamented that reverence for life, though an appropriate foundation for civilization, itself still remained ungrounded. Now, some eighty years later, we are in a position to do better, not only to ground an ethic in a worldview, but to use that grounding to understand the deep and rapidly growing tragedy of our time.

As a result of twentieth-century science, we have a better understanding of what makes life possible. In *What Is Life?* Schrödinger wrote: "The device by which an organism maintains itself at a fairly high level of orderliness (= fairly low level of entropy) really consists in continually sucking orderliness from its environment."[19] Let's see how it works. The universe is governed by a number of physical laws—one highly pertinent to our common future at this time in history is the second law of thermodynamics. Overall, the universe is running toward low energy, toward *entropy*. But there are also processes that establish that order as energy is taken from one area and concentrated in another: this is *antientropic* capacity. The resulting order sometimes expresses itself in *self-organizing ability*—in plants, animals, hurricanes, ecosystems, stars, crystals, thinking, and the like. As both Chardin and Chaisson point out, evolution is a general property of the universe. It occurs both at the macro and at the molecular levels. Life on earth results from vast cosmogenic processes such as cosmic expansion, gravitational clustering, chemical evolution, radiation, and the production of planets, stars, and galaxies. And, having once been formed, the earth is affected by cosmological events such as meteor strikes, ultraviolet radiation, solar wind, and multiple other events external to it but internal to the universe. The universe is, thus, a place of continuous birthing and simultaneous decay. Our challenge is not just to see ourselves embedded in the ecosphere, a common goal in environmental ethics, but to see the *ecosphere embedded in the cosmos.* To exit, or even understand, our *debaucheries,* we must rediscover our place in the drama of the cosmos.[20]

By our hand, self-organizational capacity is going away faster than it is being built up. This is happening *now:* the orgiastic, flamboyant period. This has allowed for a temporary rise in the human population and consumption—which, in turn, is decoupling the macro- and microenvironment at an ever-increasing rate. We are reducing and undercutting the diversity and structure of ecosystems that support biodiversity, retard erosion, filter pollution, and dissipate storm surges, among other things. We are also taking a larger and larger share of the energy *flow* for ourselves as we deplete the world's forests and expand agriculture—leaving less and less for other species.[21] *Our ethics and our behavior fly directly and massively in the face of the requirements of reverence for life.*

CAN REVERENCE FOR LIFE HELP US DECIDE WHAT TO DO?

Schweitzer is often criticized for failing to provide much guidance on how to actually implement this ethic. His own decisions seem ad hoc. He fed dozens of fish to his pet pelican. He gives detailed instructions on how to get a snake who steals eggs to strangle itself on wire. So what may seem compelling at the theoretical level may appear not very practical, even hypocritical. Indeed, it is apparent that reverence for life may emphasize the lives of individuals and underemphasize our obligations to ecosystems, an emphasis rebalanced, correctly in my view, by Leopold in his ecologically based land ethic.

In formulating a response to the charge of indefiniteness, let us begin by considering an analogy with pacifism. The pacifist is often confronted by the critic with a dilemma: kill the men who are threatening your citizens, or let the citizens be killed. To this the pacifist replies "I am not primarily concerned with what to do about these dilemmas when they arise, but I am concerned with preventing them from arising to begin with. So my emphasis is on disarmament, institutions such as a world court, and opting out of the war system." Reverence for life is an admonition to live life so as to avoid, or at least reduce, dilemmas of choosing life over life:

> Standing, as he does, with the whole body of living creatures under the law of this dilemma . . . in the will-to-live, man comes again and again into the position of being able to preserve his own life and life generally only at the cost of other life. If he has been touched by the ethic of Reverence for Life, he injures and destroys life only under a necessity, which he cannot avoid, and never from thoughtlessness. So far as he is a free man he uses every opportunity of tasting the blessedness of being to assist life and avert it from suffering and destruction.[22]

Reverence for life can help us with both personal and institutional decisions. It is a lived philosophy "with calluses on its hands"[23]—and, I would add, on its soul. Indeed, it is robust enough to steer the modern quest for knowledge away from increasing ignorance. Here is how.

REVERENCE FOR LIFE: PERSONAL OPPORTUNITIES

Eating Meat

There are a number of ways we can live from this point of view. One of them is to eat little or no meat. Not only does meat eating involve killing the animal in question, but modern methods of meat production typically involve the feeding of large amounts of grain to the animals. The grain production requires monocultures that eliminate the living space of other species and require pesticides and fertilizer and other production techniques that are harmful, often fatal, to microorganisms and wildlife. Over 50 percent of the grain grown in North America is fed to livestock. Imagine the flourishing of life that would take place if this practice was discontinued or even substantially reduced. Then there is the life of the animal itself. Confinement is often such that the animal cannot turn around or even touch the earth or fellow creatures. Emotions such as those concerning sexuality, aggression, and companionship are denied normal expression.[24] When we do eat meat, it should be in small amounts and only on occasion. It should come from animals raised in humane conditions and fed from sources that are compatible with resilient and healthy ecosystems.

Energy

The practices that surround energy consumption also afford us ways of avoiding dilemmas. As already noted, it is predicted that global warming *alone* will bring about the loss of hundreds of thousands of species in the next century and further harm to the already radically disadvantaged human poor. We can avoid, or at least reduce the horror of, this outcome by drastically reducing our energy use—given the current technological mix, by reinforcing existing trends toward smaller families and accelerating the development of alternative technologies.

On the Land

In managing my tree farm in Quebec, I have been guided by Leopold's practices at his legendary farm in Sand County, Wisconsin. We should manage not toward a single objective—but toward health of the land itself. Here is how Leopold approached the issue of land health. He suggested that the symptoms of land sickness are "abnormal erosion, abnormal intensity of floods, decline of yields in crops and forests, decline in carrying capacity . . . , a general tendency toward the shortening of species lists and of food chains, and a world wide dominance of plant and animal weeds."[25] A promising way of giving greater specificity to Schweitzer's concept of reverence for life is to connect it to the idea of land health—where *land* is understood to include life and its support systems such as air, soil, and water.

My fields are farmed organically and in a manner that reduces erosion. The tree plantations, which are small and interspersed with managed and unmanaged native forest, are planted with soft woods to maximize production. Part of the yield will be harvested about twenty years after planting for pulp and the balance forty to fifty years after planting for saw timber. The farm includes native woodland managed to preserve a diversity of indigenous species so that, in the event of a major perturbation, such as climate change, major storms, fire, or disease, there will be a variety of genetic material to draw on. Finally, there are large sections of old-growth forest that are left without intervention except to remove trees that threaten diversity. For example, large poplars can overtop the rest of the forest and retard or eliminate the development of other species. This part of my farm is analogous to what

Leopold referred to as his "mighty fortress." This seems likely to be a reference to Martin Luther's hymn of the same name—and refers to a place on Leopold's farm where the processes of evolution can continue. *Its* kingdom is forever.

I am not the *owner* of this property in any strong sense of the word—but rather its temporary caretaker. I am trying to create an open future for it and its future stewards. It should be at least moderately adaptive to changing economic conditions and, depending on the speed and magnitude of climate change, to future "natural" conditions. I am aiming at what Gunderson and Holling call "ecosystem resilience." Ecosystem resilience "is measured by the magnitude of disturbance that can be absorbed before the system changes its structure by changing the variables and processes that control *behavior.*"[26] I believe that this is an example of appropriate expression of reverence for life because it results in the continuation and expansion of the diversity of human and natural systems.

Putting Civilization on an Ethical Foundation

Once we begin with a reverence for life, we can see that the question is no longer what is good for man but what is good for the flourishing of life. The urgently needed redirection calls for reconsidering what knowledge is for and a redefinition of some of the basic terms of ethics so that issues of fairness to other species may be considered.

We Need to Reenvision What Knowledge Is for and Its Relation to Ignorance

Knowledge must be grounded in service to life. The ethics entwining the sciences should be that of reverence for life. We should begin with the perspective that the earth is part of the cosmos and, accepting that perspective, define the place of human activity within it. Beginning with reverence for life as the ethical basis for science reduces, *but does not wholly eliminate,* the ways in which the knowledge-as-power paradigm produces ignorance and creates a dysfunctional relationship with life and the world. First, ignorance of its own ethical, metaphysical, and theological assumptions is reduced. We would begin with an explicit ethic and worldview that justify the scientific enterprise and, ironically, are consistent with science itself. Second, the race of the Red Queen would become less frenzied as we aim to work with life's and nature's processes, as opposed to orthogonally and indifferently with respect to these processes. Third, a legitimating narrative is supplied consistent with the findings of science itself, rather than allowing them to remain prisoners of the metaphysical and theological narratives from our Judeo-Christian heritage and the currently unraveling Enlightenment tradition. Fourth, starting with an evolutionary perspective is a big advantage to begin with. It means that the complex-adaptive-systems view of the world is built in from the beginning. While the abstractions in which we inevitably think will still often be misleading, we are forewarned of their limitations—and regard them more as approximations than as truth. Fifth, we should have a different and deferential stance toward the successful traditional cultures that have lived successfully and with respect toward life. Finally, evolution, a wellspring of life itself, is something to be venerated, not discarded. We should see our knowledge as a tiny element in a learning universe.

We Need a Principle of Interspecies Fairness

This has at least two dimensions. One of them concerns ecosystems, the other the treatment of individuals. Mankind now consumes a large percentage of the products of terrestrial photosynthesis, and our wastes affect, and often afflict, all or nearly all other species. As we are only one species out of 13–15 million, this is palpably unfair. The present ad hoc system of "nature preserves" fails any reasonable test of fairness for at least four reasons. First, the amount of land and water preserved is far too small given the right of other life to flourish. Second, setting up the reserves' boundaries legitimates irresponsible exploitation of the remaining unprotected land and water. Third, since these reserves are typically isolated from one another, many of the species that reside in there are radically at risk from exogenous events like climate change because migration between these reserves is not feasible in many cases. Finally, these reserves offer little protection against molecular imperialism.

Industrial animal agriculture involving confinement, crowding, and other disrespectful practices is incompatible with any minimal account of duties to other species. It should be made illegal.

EXITING A FAILED PHILOSOPHY OF HISTORY

I have tried to show how we can escape from an agenda of knowledge as power that seeks to control the earth and the life on it. I offer a way to escape from an understanding of history and our place in it that is proving to be not only ungrounded but dysfunctional. Here is how. In *Reinventing Eden,* Carolyn Merchant argues that our culture is trapped in a linear account of expulsion and redemption, structurally based on the story of the Fall in the Book of Genesis. She argues that even the parts of our culture that have non-Judeo-Christian roots, such as those coming from Greece, exhibit this structure. For her, the mainstream culture is attempting to engineer a return to Eden, working with the Baconian agenda of transforming and mastering nature. We seek a return to a blissful state through science, technology, and capitalism. This goal manifests itself in a variety of ways, from the control of disease to the perfect suburban house surrounded by its own perfect garden. Many streams of the environmental movement such as the wilderness philosophy of John Muir—who saw nature as perfect the way God created it—still essentially participate in the mainstream narrative by wishing to reverse it, seeking to protect, and, where necessary, restore, nature to its original pristine condition. This gives rise to an agenda that is very damaging to persons and cultures that have been living in nature, as they must now be expelled from the garden to ensure its purity.

Reverence for life does not participate in the drama of reversal from either direction. It does not seek to dominate nature, subordinating it to our will, nor does it aim to return to an out-of-reach pristine condition. It is an ethic of respect for life as we find it on the planet, yet at the same time it looks forward to being open and keeping robust the things that make life possible. As I have tried to demonstrate, it is one of a cluster of ethical conceptions now coming to the fore. Merchant urges an ethic of partnership, where we try to work along with natural processes; Aldo Leopold an ethic of humanity, where we are conscious of ourselves as members of the biotic community. I refer to an ethic for the commonwealth of life. These ideas bear a family resemblance to each other and offer signposts to, to use Wallace Stegner's phrase, "a geography of hope."

CONCLUSION

We are neither the chosen species nor the chosen people. Our minds are not inherently the clear light of reason stipulated by the Enlightenment. The Baconian project was a misstep along the way. Much was learned, but too much was cast aside in trying to make the living world simple and controllable. We do not own the world but are simply voyagers on it along with millions of other species—many extinct, many yet to come—with whom we share both heritage and destiny. *Ignorance* is another name for *humility,* humility before the mysteries of life and the universe. Humility is an ethic that can guide civilization as it charts its course through the waters of the future. Integrity is to live with grace and compassion from this point of view.

Though there is no reason to eschew knowledge, there is every reason to recognize that its current subordination to an ethics of power is such that it often generates its own opposite. In this way, it denies us what we need to live in a world where life prospers. Ironically, a more humble conceptualization of who we are and where we came from will set us free to participate in the flourishing of life and its possibilities on this planet. We are more likely to embrace that *in which we live and move and have our being* if we start with how little we know, recognize that the pursuit of more knowledge *as power* has run its course, and begin with an ethic of reverence. If we take the river as it comes, as do my Cree colleagues, we are more likely to reach the sea. We may, to use Thoreau's words, "walk even with the builder of the universe"[27] and live in harmony with a learning universe. There is no greater power, no liberty more complete.

NOTES

I am indebted to Naomi Arbit, Gilles Caron, Herman Daly, Margaret Anne Forrest, Elisabeth Fraser, Nicole Klenk, Jessica Labreque, Greg Mikkelson, Suzanne Moore, Bryan Norton, Conn Nugent, Philip Osano, Cartter Patten, Colin Scott, Bill Vitek, and two anonymous reviewers for thoughtful critiques of this essay. David K. Goodin and Jeremy Schmidt gave especially generously of their time and insights. I was substantially influenced by Harvey Feit's writings on the Cree. Defects are my own. I am grateful to the Social Science and Humanities Research Council of Canada for supporting my research and for supporting the project on Protected Area Creation, Culture and Development at the Cree Community of Wemindji, James Bay, Quebec.

1. Bacon quoted in Carolyn Merchant, *Reinventing Eden: The Fate of Nature in Western Culture* (New York: Routledge, 2004), 75.

2. To quote an old Archer Daniels Midland ad campaign.

3. Robert K. Faulkner, *Francis Bacon and the Project of Progress* (Lanham, MD: Rowman & Littlefield, 1993), 238–39.

4. John Gowdy, ed., *Limited Wants, Unlimited Means: A Reader on*

Hunter-Gatherer Economics and the Environment (Washington, DC: Island, 1998).

5. I am indebted to Philip Osano for this point about investments in science and technology related to health.

6. Mike Crang, *Cultural Geography* (New York: Routledge, 1998), 109.

7. Donald Redfield Griffin, *Animal Minds* (Chicago: University of Chicago Press, 1992).

8. George Lakoff and Mark Johnson, *Philosophy in the Flesh: The Embodied Mind and Its Challenge to Western Thought* (New York: Basic, 1999), 554.

9. I am indebted to Steve Maguire for this point.

10. Lakoff and Johnson, *Philosophy in the Flesh*, 554–55.

11. Ulrich Beck, *Risk Society: Towards a New Modernity* (London: Sage, 1992).

12. Malte Faber, Reiner Manstetten, and John Proops, *Ecological Economics: Concepts and Methods* (Brookfield, VT: Edward Elgar, 1996), 218. Much of this section of my essay relies on this volume.

13. Ibid., 220.

14. Colin Scott, "Science for the West, Myth for the Rest? The Case of James Bay Cree Knowledge Construction," in *Naked Science: Anthropological Inquiries into Boundaries, Power and Knowledge,* ed. Laura Nader (New York: Routledge, 1996), 69–86.

15. Pierre Teilhard de Chardin, *The Phenomenon of Man* (London: Collins, 1959).

16. Eric Chaisson, *Epic of Evolution: Seven Ages of the Cosmos* (New York: Columbia University Press, 2006), 436.

17. Albert Schweitzer, *Out of My Life and Thought* (New York: Henry Holt, 1933), 185–86.

18. For a fuller explanation of Schweitzer's reasoning, see my "Are There Any Natural Resources?" *Politics and the Life Sciences* 23, no. 1 (2005): 11–20.

19. Erwin Schrödinger, *What Is Life? The Physical Aspect of the Living Cell* (Cambridge: Cambridge University Press, 1945), 73.

20. This is not an endorsement of expensive theatrics in space. Much of what we need to learn can be done with existing earth-based technologies, remote control robotics, and space-based telescopes.

21. P. M. Vitousek et al., "Human Appropriation of the Products of Photosynthesis," *Bioscience* 36, no. 6 (1986): 368–73.

22. Schweitzer, *Out of My Life and Thought,* 272.

23. Marvin Meyer, "Affirming Reverence for Life," in *Reverence for Life: The Ethics of Albert Schweitzer for the Twenty-first Century,* ed. Marvin Meyer and Kurt Bergel (Syracuse, NY: Syracuse University Press, 2002), 35 n. 49.

24. For a discussion of the ethics and practice of confinement agriculture, see Matthew Scully, *Dominion: The Power of Man, the Suffering of Animals and the Call to Mercy* (New York: St. Martin's, 2002).

25. Aldo Leopold, "The Land-Health Concept and Conservation," in *For the Health of the Land: Previously Unpublished Essays and Other Writings,* ed. J. Baird Callicott and Eric T. Freyfogle (Washington, DC: Island, 1999), 219.

26. Lance H. Gunderson and C. S. Holling, *Panarchy: Understanding Transformations in Human and Natural Systems* (Washington, DC: Island, 2002), 28.

27. Henry David Thoreau, *Walden and Resistance to Civil Government, Second Edition—A Norton Critical Edition,* ed. William Rossi (New York: Norton, 1992), 220.

THE PATH OF ENLIGHTENED IGNORANCE

Alfred North Whitehead and Ernst Mayr

Strachan Donnelley

We humans inescapably face a fundamental civic challenge: our long-term responsibilities to human communities and nature in all their complex, historical, and value-laden interactions. This is a dominant moral and practical problem for which we are culturally ill prepared. This volume and the original Ignorance-Based Worldview Conference explore the proposition that our best chance for successfully meeting our obligations to humans and nature is through "the way of ignorance," that is, by owning up to what we do not know and perhaps in principle can never know. By such admission, we might at least avoid the blinding and dangerous hubris of claiming to know (and to control) what in fact we do not know. We would gain on the side of humility, flexibility, curiosity, and caution. Warning against such hubris has historically been at the core of philosophy, going back at least to Socrates and the Greeks.

Here, I want to examine the way of ignorance in a particularly modern setting. I want to discuss two seminal twentieth-century thinkers: the mathematician-physicist-philosopher Alfred North Whitehead and the evolutionary biologist–philosopher Ernst Mayr. Their thinking bears directly and deeply on our cardinal issue of the interplay of knowledge, ignorance, and moral action.

Whatever their differences, these two scientist-philosophers share much in common. They are both centrally interested in the nature and ultimate worldly significance of organic, including human, life. They both persuasively argue that dominant ideas inherited from the Western tradition block an adequate understanding and appreciation of organic and human life, including our civic and moral responsibilities to the earth and its residents. Both agree that, in overcoming the obstacles of the tradition, we in particular must renounce an abiding and pernicious cultural habit: the quest for certain, unassailable, final truth. (At bottom, this is the problem of dogmatism of all stripes.) The dream of certainty and final truth has animated Western thinkers and cultural communities from the early Greek Presocratics to today. We could quickly rattle off a notable philosophic all-star team: the Greeks Parmenides, Pythagoras, and Plato (in some of his moods); Descartes, Spinoza, and Leibniz of the seventeenth-century Age of Genius; Hegel and other nineteenth-century idealists; and, finally, the twentieth-century hangovers, the positivists, who will countenance only "indubitable" knowledge. But what do thinkers have in the absence of this immemorial goal, the quest for certainty and final dogma? Arguably, we have, as an alternative, the way or path of enlightened ignorance. By *enlightened ignorance,* I mean accepting what limited and circumscribed knowledge we do have but recognizing explicitly that this knowledge is fallible, subject to revision, and interwoven with much that we do not know (our ignorance). How this basic human condition plays into the question of moral judgment and action is the question at hand and constitutes the relevance of Whitehead and Mayr.

WHITEHEAD'S NATURE LIFELESS, NATURE ALIVE

According to Hans Jonas, himself a noted ethicist and philosopher of organic life, Whitehead, as an antidote to the failed tradition, was perhaps the most important philosopher of the twentieth century. Whitehead's philosophic cosmology or philosophy of organism is a major philosophic mountain, difficult to ascend and even more difficult to descend, that is, to explain to others. Characteristically, it is avoided by most philosophers, if not more courageous (or rash) theologians, whether because of its difficulty or because of its being out of contemporary fashion. But what are these philosophers and the rest of us missing? Who is truly out of step and style? And why?[1]

Whitehead's philosophic vision is highly original and is best approached with some care. At his most systematic (*Process and Reality,* 1929), Whitehead offers a general scheme of fundamental ideas in terms of which to interpret or understand the full range of human experience. He considers such philosophy our birthright, the romance of rational thought, the speculative attempt to make sense of ourselves and our world, including the many-leveled values and importance of earthly existence. Whitehead's endeavor is crucially animated by an underlying faith: the world or cosmos is coherent; things hang together, make sense, have their own significance, and can be rendered more or less intelligible.

Whitehead's method of philosophy involves both an empirical, critical exploration of human experience and an exercise of rational imagination. There are rigorous demands of coherence and consistency (rational demands), plus adequacy to experience and enlightened observation, that is, an ongoing check against our experience of ourselves and the world (empirical demands).[2] Yet, even at philosophy's best, there is, Whitehead claims, no guarantee of success or certainty. Philosophy and our explorations of human experience are a never-ending adventure, a critical and continuing revolt against all cultural, intellectual, and metaphysical pretension and dogmatism. (Here, at the outset, note Whitehead's contention of the fundamental limitations in all ways of knowing. Certainty is a ruse. Certain knowledge is beyond our human, intellectual powers, logic and mathematics notwithstanding.) Yet he further claims that a philosophic worldview, formal or informal, is a presupposed background of all human thought, experience, and action. There is no escaping such fundamental frameworks of thought and feeling. Whether acknowledged or not, we are all by necessity "world-viewers." Thus, philosophy (and ethics) and the way of uncertainty or enlightened ignorance must ongoingly learn to travel hand in hand in their never-ending openness to the world.

This is Whitehead's neoclassical, grand-style approach to philosophy, minus the traditional quest for certainty and final truth. As such, it cuts against the grain of modern positivist sensibilities, which mean to stick to indubitable facts and avoid all metaphysical nonsense or speculation, resolutely ignoring all that is dubious.[3]

Whitehead counters that this ascetic, if not arrogant, act of positivist self-deception is precisely the problem. To make his point, he undertakes a critical survey of modern science and philosophy begun in earnest in the sixteenth and seventeenth centuries.[4] Whitehead is specifically interested in the reigning, dominant worldview that he terms *Common Sense* or *Sense Perception Nature.* This modern science and philosophy of nature rests on a particular theory of knowledge (epistemology): a reliance on *sense perception,* especially "clear-sighted" vision, as the sole legitimate mode of experience, observation, and knowledge of the world.[5] Such a conception of nature does embody general and practically important truth. But, beyond its limited scope, it generates a complete muddle that philosophically acquiesces in the unintelligible, the incoherent, and the meaningless. It thereby undermines civilized thought and action, morally practical and other. This is Whitehead's core, radical indictment of our modern culture.

Whitehead responds to this pervasive doctrine by underscoring the inadequacy, limitations, and relative superficiality of sense perception: mere sensa (colors, sounds, tastes, touches), spatially or temporally distributed, that tell no tales of themselves, their origins, or their changes. In our experience, sensa are just immediately present in consciousness. End of story. Whitehead further notes the curious hybrid character of sense perception: its twin origination in the physiological functionings of the organic body (eyes, ears, skin, and more) and our extensive, geometric, spatiotemporal implication in the evolving world. (The world is "extensively" arrayed before our experiencing selves.)[6] Moreover, we do not need professional philosophers to point out the occasional and recurrent illusions of sense perception, for example, puddles of water on shimmering summer highways. That sense perception is the terra firma of modern science and philosophic cosmology strikes Whitehead as curiously and decidedly odd.

Why would the modern philosophic tradition place such a heavy reliance on sense perception? There are good historical reasons. For one, it closely aligns with positivism, with its desire for uncontestable experiential facts as the basis of all empirical, worldly knowledge and its (often covert) quest for certainty. Whitehead corners his quarry, his bête noir, and mounts a counterattack. Positivism acquiesces in ultimate irrationality and incoherence in the name of the observationally and conceptually indubitable: unquestionable, clearly observed, hardheaded facts and correlative ideas. The modern philosophic forebear of this mood and doctrine is the seventeenth-century figure René Descartes, with his clear and distinct ideas, guaranteed as certain and true by God and his goodness. From Whitehead's critical perspective, positivism fails resolutely to press for connections (relations) among things and, thus, for any serious philosophic intelligibility. (For Whitehead, making connections between things or ideas—"connecting the dots"—is the heart or essence of rational thought, not the narrow string of deductions of logic or mathematical proof.) Specifically, positivism glaringly fails to link nature, life, mentality, and value, which we emphatically feel to be intimately connected in everyday, prephilosophic human experience.

Whitehead fastens specifically on the historical demise of modern science's Sense Perception conception of nature and its lingering philosophic hangovers, a tale worth retelling, if only briefly. Thanks importantly to Descartes, early modern cosmology excludes life (animate, organic life) and mind from nature.[7] Nature is conceptually reduced to permanent "bits of matter" passively supporting qualities or characters (e.g., sensa) in an empty, geometric (three-dimensional) space. Locomotion, change of place, "billiard-balls-in-motion," is the sole or dominant mode of change.[8] All such material things are *simply* or *singly located.* They just are where they are. Nature exists *in the instant,* that is, atemporally, with no essential reference to time. There are no fundamental relations of individual things to other individual things at other times and places. This is nature as dominantly disclosed in and by sense perception, especially vision.

Sense Perception Nature straightaway yields derivative philosophic puzzles or quandaries. These puzzles go beyond the fundamental incoherence introduced by Descartes' mind-body dualism: the radical split of mind (*res cogitans*) and matter (*res extensa*), which have no real, essential, or ontological relations to one another. As noted by both Whitehead and Jonas, such a split-world doctrine leaves no room for recognizing and understanding organic life. (Organisms are decidedly not just bodiless minds or "bits of matter," physical automatons, or subtle natural machines.) Moreover, in the modern era, transmission theories of light and sound led straightaway to the *subjectification* of sensa, which we humans naively or straightforwardly experience to inhere in nature—the colors, sounds, tastes, smells, and tactile qualities of things.[9] Nature is denuded and devalued. All secondary (nongeometric or nonextensive), aesthetic, sensory qualities are banished to the anatural, Cartesian mind, leaving nature poets unknowingly, quips Whitehead, singing hosannas only to themselves. In sum, sense perception and its relations to the world abroad become highly problematic, if not unintelligible, in this conception of nature.

On the physical, material side of the dualistic divide, things fare no better. There is no intelligible reason for the gravitational stresses among Isaac Newton's spatially distributed, simply located, "massy" bits of matter. True, we have mathematical formulas about nature's ways, but in themselves such formulas give no philosophic reasons.[10] Why the world's geometric extensiveness? Why the massiness of matter? Why the gravitational stresses? For Sense Perception Nature and its animating positivist spirit, these are mere brute facts with no intelligible connections or compresence.

Intellectual progress since the seventeenth century has effected the *scientific,* if not the underlying philosophic or conceptual, demise of the Common Sense or Sense Perception conception of nature, with its empty space, permanents bits of matter, and the later (eighteenth to nineteenth century) halfway house of a jellylike ether that accommodates all sorts of intracosmic stresses and strains. Scientifically, there has been the complete denial of nature as simply located and existing at the instant.[11] In its place has emerged Nature energetic, enduring, becoming on all scales (cosmic or regional) dynamic and historical. This constitutes a fundamental sea change in thought. Nature is one grand theater of incessant activity with both enduring and changing patterns and habits of behavior among "group agitations" (the old bits of matter). There are cosmic epochs of energetic activity with different cosmological characters, for example, geometric patterns and electromagnetism. However, the characteristics carry no mark of absolute necessity, as claimed by old ironclad laws of nature. (Unchanging "laws of nature" nicely go hand in hand with belief in atemporal [unchanging], certain truth.) But, Whitehead asks, why? Why the activity? What is getting effected? Is this a cosmic theater of the absurd—a meaningless sound and fury signifying nothing? Has physics been reduced, or has it reduced itself, to a "mystical chant over an unintelligible universe"?[12]

On these questions, physicalist science's positivist spirit remains silent and sticks to its guns, sense perceptible facts and their interpretation, mathematical or no. Here, according to Whitehead, is the fateful philosophic result of a Dead Nature that yields and can yield no reasons.[13] Whitehead boldly breaks from the pack and recoils from such a flat, lifeless, unintelligible conception of nature. Having critically and historically exposed the emperor's clothes—the philosophic bankruptcy of positivism—Whitehead opts for a speculative conception of Nature Alive. Indeed, Whitehead considered the status of life in (positivist, physicalist) nature to be *the* modern problem of philosophy and science. "The very meaning of life is in doubt."[14] This indeed is a radical charge. If life is in doubt, then we are in doubt as to our meaning and status, for, first and foremost, we are human organisms living in the world. We may add, some seventy years later, that the practical status of life in nature is perhaps *the* long-term problem of the political and civic world. Consider the present and ongoing biodiversity crisis, as connected with habitat destruction and climate change or global warming. The theoretical and practical problems are not unrelated.

Enter Whitehead the modern classical speculative philosopher who refuses to accept ultimate irrationality, absurdity, and meaninglessness. In particular, he rejects the disconnection of fundamental and abiding details of our human experience of nature, life, mind, and value. He does so by philosophically and resolutely fusing nature and life. Interestingly, it is here that Whitehead himself most explicitly and directly turns to a path of enlightened ignorance. He turns his attention to life, living nature, at the cost of any pretension to final philosophic certainty.

Whitehead requestions human experience, including the putative clarity and importance of sense perception. Again, note the pervasive, though often silent, reference of sense perception to the organic body and its sense organs—eyes, ears, fingers, nose, and tongue. Don't be seduced by the intermittently clear sensa and the conceptual abstractions that we build on them and call reality, absurd or no. This is to commit the fallacy of misplaced concreteness, mistaking the abstract or partial for the concrete in its full complexity.[15] Attend instead to the more fundamental, imperative, vague, and incessant feelings of the organic body, which are anything but "clear and distinct" and are captured only "through a glass darkly." There you will find the immediacy of life, emotional and active, and our personal, intimate human implication in our bodies and the world abroad—nature, life, and mental functionings experienced together. For Whitehead, these are the fundamental facts or deliverances of experience that philosophy ought to connect and render intelligible, rescuing the meaning, value, and significance of nature, life, and ourselves by *speculatively* interpreting their interconnections, their status vis-à-vis one another, the *reasons* for their compresence. Giving such reasons is not the same as claiming certain truth.

Herein lies Whitehead's originality and speculative boldness, his decisive break from modern tradition. He stares down Descartes, Hume, Newton, Kant, and other typically skeptical moderns, all in the name of philosophic sanity and civilized practical and moral activity, again with no claim to dogmatic final certainty, which no discipline enjoys, whether mathematics, physics, philosophy, theology, ethics, or some other. The way of enlightened ignorance, fallible knowledge and ignorance intermingled, is an adventure of ideas, undertaken with a combination of boldness and humility.

Forsaking rationalist meditation (Platonic and Cartesian), and echoing more earthbound naturalists (e.g., the Presocratic Heraclitus), Whitehead digs deep into personal bodily life and experience.[16] Here he finds a certain *absoluteness of self-enjoyment,* an emotional appropriation of the antecedent functionings of the organic body and the physical universe (nature) on the part of an immediate, subjectively alive "occasion of experience." Here he finds an episode of *creative activity,* which is precisely this appropriation, with its mating of the actual or the already realized and the potential, the never before realized, in a new pattern of individual, emotional, value-laden world experience. Here he finds the *efficacy of aim,* involving mental functions more or less effective and dominant, which guides the process of appropriation and self-creation, always with one eye on the past and the other on the impending future. Whitehead philosophically generalizes from this personal, human, concrete bodily experience of the world. (As an integral part of the world ourselves, we are clues to the universe abroad.) Here are nature, life, value, individuality, and mentality together in the final, really real things (actualities) of the world, including their creative implications in the advancing natural universe.[17] The natural world is decidedly more than the "sheer or mere activities" studied by the reductive physical sciences, with their abstractions from the naturally concrete. The more concrete or comprehensive perspective of speculative philosophic cosmology unearths Nature Alive, either immediate or past, the scene of multiple values, novelly emerging order, and intensities of experience and existence aimed at, realized, or frustrated. Here is a worldview that is a far cry from the lifeless, bloodless, and unintelligible universe of Nature Dead and the corresponding private fantasies of aworldly and anatural human minds, the legacy of Descartes and the seventeenth century, which still fatefully haunts us. We are philosophically reintroduced to the world in which we live.

Whitehead's bold speculations may be a romance of rational philosophic thought. But they are not idle musings. Though renouncing any claim to final truth (we, as noted, are finally *Ignoramus*), Whitehead presses hard for philosophic intelligibility and for giving the world and our human selves their experiential due. Whitehead fuses nature, life, and mentality in interconnected, worldly episodes of subjective experience and existence. (These episodes of experience arise and perish in constituting the creative advance of the universe.) In so doing, he philosophically explains or interprets what the worldview of Common Sense or Sense Perception Nature cannot. Whitehead intelligibly interprets the experienced factors of our world: real causation or causal efficacy (both efficient and final or goal directed); memory (the past effective in the present); why definitely charactered things and occasions of experience must arise where and when they do (dynamic, historical spatiotemporal contexts as fundamental); why Instant Nature is an obfuscating fiction, perhaps mirroring an eternal and unchanging transcendent reality, itself suspect; why endurance, becoming, and change—interconnections of process and reality—are fundamental aspects of ourselves and the world within which we live; why value experience and existence, human and nonhuman, are fundamental and widely shared and are pervasively experienced as such, solipsistic philosophies and philosophers notwithstanding.[18] Whitehead claims that our fundamental experience of the world is "have care, here is something important." Moreover, for the sake of fundamental philosophic intelligibility, Whitehead speculatively posits an "Ultimate," *Creativity,* which ongoingly gathers the many occasions of experience of the past into novel, immediately alive occasions of experience, thus giving Nature Alive its fundamental character of an ever-becoming, creative advance into novelty.[19]

For those who dwell and toil in the vineyards of everyday life and take worldly responsibilities seriously, there is one central doctrine of Whitehead's Nature Alive that particularly provokes interest: his notion of identity and "mutual immanence." This is his interpretation of the fundamental togetherness or interconnectedness of the world's individuals or final "really real" things, his decisive and important move beyond atomistic cosmologies or worldviews.[20]

Starting from our immediate human selves, Whitehead notes several claims to unity and identity, personal and otherwise, based on inescapable feelings of derivation and continuity: the identity and unity of ourselves with our emotional, living bodies; the identity and unity of ourselves with our past stream of personal experiences and existence; the identity and continuity of our living bodies with the rest of dynamically functioning and evolving nature. (In the end, all things are connected.) These various identities and continuities merge and interweave in our immediate human selves: who we are now, more or less active and creative, with real ties to the past and future of our personal lives, our own bodies, and the wider world, humanly cultural and natural. We

always and fundamentally find a double sense of *inclusion*. The enduring self, the body, and the world are experientially found to be in—
within—our present experience, as the dynamic basis of our immediate,
lively, and self-creative activity. *The world is in the self.* Alternatively,
we find our immediate selves actively embedded within our enduring personal lives, living bodies, and historically cultural and natural
worlds. *The self is in the world.* Here is Whitehead's doctrine of *mutual
immanence*—the double inclusion of self and world—that is meant to
recognize and philosophically interpret a central fact of our experience
and existence: the essential togetherness of self and world. The self and
the world are *internally related;* that is, they essentially matter to one
another with respect to both existence and character (particular forms
and capacities). (Interestingly, Whitehead here was probably influenced
more by advances in physics than advances in evolutionary biology. In
"field theory," energetic parts are in the whole, and the energetic whole
is in the parts.)[21]

"Mutual immanence" is the heart and soul of Whitehead's philosophic cosmology of Nature Alive, a doctrine conspicuously absent
from the cosmological interpretation of Nature Lifeless or Dead (physical, material, atomistic nature). For our purposes, we should explore
the import of mutual immanence in shedding light on and interpreting
our ethical experiences and responsibilities to human individuals and
communities and wider nature. For example, which philosophic scheme
makes better sense of and elucidates conservation philosophy and ethics, for example, the worldly aesthetic dimensions and ethical impera-
tives of Aldo Leopold's *A Sand County Almanac,* including its land
ethic? Leopold claims that the good and the bad, right and wrong, are to
be understood in terms of the protection and promotion of the dynamic
and historical integrity, stability, and beauty of biotic communities and
ecosystems (interrelated flora, fauna, and abiotic elements, from the regional to the global), which emphatically include our human selves. Is
Leopold better interpreted by the doctrine of the mutual immanence of
ourselves and the natural world of Nature Alive or by the mutual exclu-
sions, alienations, and unintelligible connections of our human selves
and Nature Lifeless or Dead? Which helps us better face a bewilderingly complex, forever changing but deeply value-laden world, natural
and cultural?

Is not Whitehead's speculative scheme, his path of enlightened ignorance, closer to the mark than a philosophic interpretation that finds

poets and other culturalists foisting humanly subjective values on an alien, ahuman nature and indulging unwittingly in self-congratulation or nervously whistling in the dark? More generally, in getting our ethical heads screwed on right, is not Whitehead profoundly right that fundamental philosophic worldviews or cosmologies, our basic interpretative schemes of thought, no matter how "finally" uncertain and open for revision, really matter (positively or negatively) with respect both to thinking and to practical action? In the end, how we think, so we act.

We ought to turn these critical remarks and questions to good use. We should further press Whitehead into philosophic service, aimed at our and nature's future. As we have seen, Whitehead claims no final truth or certainty for his speculative philosophy and interpretation of the world. Even if we have quarrels with aspects or the final adequacy of his philosophy, as he did himself, it is not thereby rendered useless. Significantly, Whitehead has reclaimed the philosophic high ground of Nature Alive and forcibly warned us against fallacies of misplaced concreteness. Beware taking our own conceptual abstractions (scientific, philosophic, and other) for the living world's complex concreteness. Whitehead gives us a crucial purchase or perspective for critically engaging all philosophies of nature, including those inspired by Darwin and Leopold. Here is an ongoing and fundamental moral and philosophic challenge, with no end in sight.

MAYR'S DARWINIAN REVOLUTION

I have briefly explored Whitehead's speculative philosophy of organism, his own path of critical, enlightened ignorance. What is Ernst Mayr's path? How do the two paths compare? What can we learn from the comparison?

Mayr was arguably the dean of twentieth-century evolutionary biology. He spent the better part of the century defending and elaborating Darwin's theory of evolution. His reasons for this lifelong personal and professional passion are clear. Mayr considers the publication of Darwin's *The Origin of Species* (1859) to have inaugurated the single most profound scientific and philosophic revolution in the history of Western thought. Indeed, he considers Darwinism so far-reaching as to move well beyond scientific biology and to constitute a genuinely new philosophic and moral worldview (in Whitehead's terms, a speculative cosmology).[22]

Fundamental presuppositions of the tradition are challenged, and we are led into new philosophic and ethical territory, whether or not most of us have followed or digested this venture.

The core new ideas established by Darwin and his followers are the common descent of all life (species and individual organisms) from a single historical origin, coupled with life's dynamic proliferation and diversification by an evolutionary two-step: phenotypical and genetic variation (genetic mutation and sexual recombination) and natural (and sexual) selection or elimination. Here is a dynamic, historical process whereby generations of individuals, populations, and species of organisms become gradually adapted to changing environments and new ecological niches via differential survival of individuals (and, perhaps, species populations) and reproduction.[23]

According to Mayr, to understand the theory (indeed, the by now well-established fact) of Darwinian evolution, we must discard and move well beyond fundamental and long-engrained tenets of Western thought. Mayr explicitly points to cosmic teleology; physicalist, Newtonian determinism; and essentialist or typological thinking.[24] As we shall see, rejection of these three tenets in combination amounts to moving to another path of enlightened ignorance: science and philosophy basing themselves on a new bedrock of "uncertain" knowledge (fallible knowledge amid ignorance).

Cosmic teleology is the doctrine of the grand design of the cosmos, including earthly life, by a divine, intelligent, and purposive designer or God. In the newly understood scheme of things, over geologic, ecological, and evolutionary time, nature creates in passing its own organic forms, capacities, and order.

Similarly, the old traditional notion of billiard-balls-in-motion causation, material antecedents strictly determining material consequents, the hegemony of efficient causation so crucial to Newtonian and Cartesian science, is gone. Rather, there are innumerable causal influences at work on many spatiotemporal scales. Moreover, historical contingencies and chance have a real role in concrete worldly outcomes. Further, according to Mayr, there are two modes of organic causation that must be taken into account, ultimate and proximate or physiological causation.[25] In the new science and worldview, *ultimate causation* refers to the significant causal influence of historically engendered genomes in the development and behavior of organisms. Genomes and genetic material did not come under the purview of traditional Newtonian physicalists and, indeed, constitute, or should constitute, a genuine philosophic puzzle for any materialist, Mayr included. How can matter take on such form-engendering and directing powers as genomes have, notwithstanding all the environmental interactions required for the realization of phenotypical, or individual organismal, results? A good question worth pondering. Whatever, Newton's strict determinism is out, and infinitely more complex, less deterministic and predictable "orchestral causation" is in.

Let us briefly consider orchestral causation, for it is a key new conception or metaphor for the time-honored notion of causation. Think about the performance of an orchestral piece of music, say Verdi's *Requiem.* Who or what is the cause of the performance? Is it Verdi, the score of the *Requiem,* the conductor, the members of the orchestra and chorus, the soloists, the acoustics of the orchestra hall, and the "musical ears" of the audience, among other factors? Seemingly, there are no single or singly sufficient causes. Rather, the result is the outcome of the *interactions* of all the contributing factors. Change any of the constituents, say, the musical abilities (or moods) of the orchestra and chorus, and the results will be different. (Old Heraclitus was right. You cannot step into the same river or world of becoming twice.) Orchestral causation, and not "billiard-balls-in-motion" causation, seems fundamentally to characterize the earthly realm of life, if not the natural universe as a whole.

To return to Mayr's tripartite critique, Darwinian perspectives render essentialist or typological thinking otiose.[26] In reality, there are no species types—dog, rose, fish, human being—with differences among individuals of any species being merely accidental or adventitious and unimportant. Rather, for Darwinians, there are only individual organisms with all their differences, phenotypical and genomic. Moreover, these differences make all the difference in evolutionary, ecological history. Without individual differences, natural selection would have nothing to select. "Adaptational life" could not historically carry on, or, rather, it would never get going in the first place.

The demise of essentialist, typological thinking means the correlative ascent of "populational thinking."[27] There are only populations or communities of individually differing organisms. If these individuals actually or potentially interbreed, they belong to the same species. (This is evolutionary biology's most prevalent definition of species.) If they do not, they belong to some wider, interactive community of ecosystemic life.

Combine the notions of dynamic orchestral causation and popu-lational thinking, and something philosophically striking emerges: the fundamental notion and earthly phenomenon of "emergence" itself.[28] For example, individual organisms are no self-sufficient substances or "atoms." They exist only in dynamic, orchestral interaction with the world. Indeed, their very individual being—with whatever subjectiv-ity or selfhood they may enjoy—emerges from and within worldly or-chestral causation, whatever genomic contributions there may be. We humans and all other organisms are "emergent ones" and are so as long as we individually remain alive. Here is a philosophic sea change or revolution of the first order.

Here, we might take another pause and connect explicitly the Dar-winian overthrow of the tradition with the path or way of enlightened ignorance. Consider specifically the old and abiding nemesis: the quest for certainty, for certain, indubitable, and unchanging or eternal truths. The traditional assumptions inspire, or are inspired by, such a quest. (Here is an interesting philosophic chicken-and-egg problem. Which fundamental ideas inspire other fundamental ideas, or do they, in a demand for final coherence, mutually inspire or lead to one another?) There is an omniscient, eternal, and good God who directs the fashion-ing of the world (cosmic teleology). There are materially deterministic "laws of nature" that brook no interferences (physicalist determinism). There are unchanging essences of all things of all types. God, laws, essences: the quest for certainty seems the only plausible originating animus or outcome. Demolish these three pillar assumptions, and the Humpty Dumpty of the quest for certainty comes tumbling down. Amid the debris, we are left on the path of ignorance, equipped only with finite capacities for experience and knowing evolved by nature and our historical human cultures.

Mayr warmly embraces this decidedly human path of knowing. He has his own nontraditional and biologically inspired tenets for gaining scientific and philosophic understanding, a Darwinian epistemology or theory of knowledge.[29] Consider the bewilderingly complex, unique his-tory of the evolution of earthly life. Make use of whatever comparative observations and experiments seem appropriate. Respect evolution's plural ways and contingencies. Above all, boldly speculate; engender hypotheses or historical narratives; clearly define crucial terms of ex-planation or interpretation; and then check reflectively considered ideas back against the historical, worldly record (evidence) and take on all comers, all rival interpretations of evolutionary facts. For, in principle, there is no guarantee of certain truth, the epistemic benchmark of the old tradition. Let the most robust and adequate theories, narratives, and interpretations thrive and carry on in the ongoing battle or adventure of ideas. Unmistakably, Mayr exhorts us to follow a path of enlightened ignorance or "uncertain" truth.

WHITEHEAD AND MAYR TOGETHER

Let us return to the fundamental, double theme of individual organic being and worldly interaction. Here, Mayr's thought and Whitehead's inescapably converge or intersect. Despite thinking well in advance of the "genetic revolution" and the evolutionary synthesis of the 1930s–1950s, Whitehead incorporates in his philosophy notions consistent with orchestral causation (worldly interrelations and interactions) and emergent individual organisms, up to and including human individuals. Recall our discussion of an "occasion of experience" and the relation of self and world, including their mutual immanence. Further, Whitehead has an explicit interpretation of the complex organization and experiential capacities of biological and human organisms. The explanations are sophisticated and complex. Fortunately, we need not go into the details here, but the human individual, experient, or self in the full reaches of his or her life is understood by Whitehead to be fundamentally and on-goingly dependent on the living organic body and the biotic and aboitic, as well as cultural, world abroad.[30]

Moreover, Whiteheadian perspectives and interpretations help deepen genuine philosophic puzzles inescapably raised by organic life and, thus, Darwinian biology. Again, remember fundamental Whiteheadian tenets: the dynamic, mutual immanence of the immediate subject, the enduring self (if there be one), and the world; that worldly reality by necessity involves subjectivity, emotion, purposiveness, and, thus, value, especially the concrete value that is an actuality's or an organism's self-constituting activities, arising in response to the given world and aimed at the worldly future. In short, from Whitehead's perspective, Mayr's organisms, thanks to the ultimate causation of genomes, do not merely exhibit teleonomic or goal-directed functioning, bodily or behavioral, no matter how closed or open-ended, that is, open to worldly, including cultural, influences. Rather, organisms are real individuals, real subjects, with real lives, emotions, purposes, experiences, and more, as is appropriate to the complexity of their organic organization and worldly situation. (The purposive, goal-directed behavior of individual organisms or social communities is not to be confused with cosmic [supernatural] teleology. Capacities of purposive behavior have historically evolved.) Mayr, evolutionary biologists, indeed, all of us must beware of the fallacy of misplaced concreteness and not reduce organic feeling and valuing subjects to organic objects and functions. To do so would blunt serious philosophic reflection on the value and significance of organic life, including our own, and our moral and civic duties to worldly organic communities and individuals, human and other.

For the moment, I want to set aside these ultimate philosophic and moral matters, no matter how important. I want to revisit the cardinal issue of the origination of worldly order, cosmological, biological, and humanly cultural. Here is a philosophic issue that has fascinated humans from the beginning, from the early Greek Presocratics and before. Modern life, with its own underlying cultural and intellectual traditions, has, it is hoped, not so dulled our philosophic ears that we cannot be deeply fascinated and puzzled by the reality or very being of order, which ongoingly cries out for philosophic interpretation.

Mayr, an honest and humane naturalist and self-avowed atheist, has disowned all cosmic teleology and essentialist thinking. Evolutionary and ecological order arises through the historical, dynamic interactions of the biotic and abiotic entities of the natural world. (Mayr ascribes to a blind tinkerer rather than a divine intelligent designer or watchmaker.) New organic forms and capacities arise through genetic mutation and recombination, coupled with natural and sexual selection (interactions with the world). With this account, Mayr feels that he follows Darwin, and, moreover, such explanation is all that he needs *as a scientist.* But this should not lull us into philosophic slumber.

Here, Whitehead may genuinely help us stay philosophically wide awake. His philosophic explanation of the introduction of order into the world and the course of evolutionary and human history is somewhat at loggerheads with Mayr. Whitehead retains his own forms of the banished tradition, cosmic teleology and essentialism.

In all fairness to Whitehead and genuine philosophic reflection, we should note at once how far Whitehead himself moved away from the Western tradition to a modern organicist position. In the most systematic expression of his speculative philosophy, *Process and Reality,* Whitehead does philosophically employ a (novelly conceived) God who influences the world, its advance into novelty, and its ongoing creation of order, including the earth's biological and humanly cultural history.[31] However, Whitehead's God does not impose a grand design on the world. Rather, he persuasively lures the individual actualities of the world toward the creation of worldly order requisite for a depth and intensity of emotional, value-laden, and purposively creative experience. He does so by a divine causal contribution to creativity's transition from the world's already realized occasions of experience to the novelly immediate, self-constituting occasion, or creativity in its throes of advance into novelty. (Here, as we have seen, is Whitehead's "enlightened ignorance" or bold, uncertain speculation par excellence.) God, via helping establish the *creative aim* of the new occasion, what it might become in its experiencing the world, provides the novelties into which creativity may advance.

The novelties provided are *forms of novel order* as potentials for the new actuality and, through the actuality, for the world to come. Here is a form of essentialist thinking. However, it is but a faint echo of Plato, Descartes, and the tradition. Whitehead claims that, in one form or another, we all must accept Plato's uncreated, unchanging, eternal forms of order or character.[32] (*Eternal objects* is his technical term.) But, for Whitehead, these forms are not final realities with ultimate significance, à la Plato. Rather, they are uncreated potentials available for individual actualities in constituting themselves and, between themselves, the world. We live in a world laced with forms of order (and disorder). Forms of order must come from somewhere. Either they are radically created in and by the natural world—ostensibly Mayr's position—or they are uncreated and made available to the world in its creative advance into novelty, into new actualities and new forms of order engendered in and between themselves. For challenging philosophic as well as, perhaps, religious reasons (e.g., upholding the objectivity of worldly experience: I here and now can experience and subjectively reenact the orderly emotional experience established there and then), Whitehead opts for uncreated forms of order (potentials) felt in their mutual relevance (relatedness) by God and appetitively made available to the world via the creative aims of becoming actualities.[33]

Without further miring ourselves in his complex and characteristically systematic interpretation of God, world, and forms of order, suffice it to say that Whitehead is struggling with a genuine philosophic problem—how there can be a world evidencing order and advancing into new forms of order. Moreover, Whitehead duly recognizes that the natural (and humanly cultural) world "orchestrally" has a great deal to say about what forms can and will be realized and that actualities are free to choose among the related promptings provided by God. They have a certain circumscribed freedom—the more so if they belong to and help constitute the more complex actualities (individual epochs of becoming) of biological organisms, including human beings. (Elaborate mental functionings, the reworking of fundamental worldly experience, are key to such circumscribed freedom.)

In short, empirically, if not also philosophically, Whitehead is much closer to Mayr than to the tradition when giving due recognition to the historical dynamics of the evolutionary, ecological realm of life. He shares with Mayr a "populational" mode of thinking. Individuals in their differences and in their relations with one another are what ultimately count, have value, and influence future worldly becoming. But, for ultimate philosophic interpretations, he does not totally abandon the philosophic tradition, with its arsenal of divine metaphysical functions and uncreated, Platonic forms, "the sources of all order."

For us, this is highly instructive. Though worldviews (according to Whitehead) are to be judged relatively to one another by their rational coherence and adequacy in interpreting the full reaches of human experience, we do not have to race to the side of Mayr or Whitehead. Philosophy is not—or should not be—a heavyweight match from which one contender must emerge victorious. Indeed, given our limited powers of knowing, this might be finally impossible. We can stand before a philosophically ultimate, important, and knotty problem that stares us in the face unresolved—here, the problem of form and order, a chief ingredient of all that is good and beautiful. In truth, we find ourselves on the path of ignorance, unknowing, *Ignoramus,* but at least knowingly so.

In short, giving reasons for the advent of order is emphatically a part of a metaphysics of ignorance or an unfinished, open-ended worldview. Whitehead, for one, would agree. Certainty, as he claimed, is a ruse, an ill-begotten child of Platonic, Cartesian, Kantian, and certain religious traditions. Despite the complexities and sophistications of our worldly experiences, all is more or less in doubt. Whitehead would include our descriptions of everyday human experience or "takings" of the world and, most emphatically, his philosophic scheme, his conception of God, and more. Whitehead considered his philosophy as a whole as a proposition or a proposal to be seriously entertained and advanced on. Mount Whitehead is there for the climbing and to provoke us on to ever more adventurous and adequate explorations of worldly experience.

From the confluence of Mayr's and Whitehead's streams of thought, we should especially remain philosophically enticed and puzzled by the ultimate creative restlessness of the universe, the ground of organic and humanly organic liveliness (perhaps forever beyond our grasp); orchestral causation; emergence, including worldly emergent individuals or selves; worldly interconnections and interactions only to be caught by forms of systems thinking; order, beauty, and goodness, especially that of the organic and humanly organic realms. The latter are also manifestly mortal, finite, and vulnerable to harm and, thus, are doubly matters of ultimate concern, both of ultimate responsibilities (moral and civic, human and natural or ecological) and of the worldly locus of the humanly good life.

In closing, I would like to lure us one step farther down the path of enlightened ignorance opened up by Whitehead and Mayr. I want briefly to explore the connections of enlightened ignorance, the systemic character of the world, and ethics. As suggested by my discussion, both Whitehead and Mayr have their own modes of systems thinking, based on correlative notions of orchestral causation, emergence, and evolving worldly order. Both have their notions of natural (and cultural) "regions within regions within regions" and upward and downward causation—inclusive wholes influencing included parts, included parts influencing inclusive wholes, a worldly hierarchy of entities characteristically changing at different rates, the "inclusive" more slowly, the "included" more rapidly.[34] For both Whitehead and Mayr, this dynamic, systemic becoming historically and contingently (nondeterministically) realizes novel forms of worldly order and value, including those realized in earthly nature and its resident human communities and cultures. Moreover, both acutely realize that worldly, earthly, and human becoming can fare for better or for worse. In short, both recognize that ultimate ethical responsibilities are thrust on us whether we like it or not.

But what are our ethical, earthly responsibilities, especially given that we are confined to paths of enlightened ignorance, that is, given that we in principle do not have, and cannot have, final and certain moral truths, dogmatically fixed moral stars to guide us? Does this situation resign us to moral nihilism or, at best, aimless moral relativism, an "I'm OK, you're OK" syndrome? No. Manifestly, this is not the way of Whitehead and Mayr. This is not the inevitable outcome of the path of enlightened ignorance. Quite the opposite. By renouncing the quest for certainty and correlative dreams of perfection, we become, or ought to become, more and more wedded to the finite and vulnerable realized goodness of earthly life—all earthly life. Given what we can discern "through a glass darkly," our moral responsibilities are systemic: to earthly processes, structures, and communities of life as well as to life's interconnected individuals. Moral responsibility is naturally ecosystemic as well as humanly communal and individual. Mayr was explicit about such Leopoldian concerns.[35] Whitehead predated Leopold, but he would have readily recognized and appreciated Leopold's morally significant "biotic communities" as a genuine conceptual contribution to the adventure of ideas.

Whitehead, Mayr, and Leopold are steadfast champions of the goodness of earthly life in all its historical plurality, biological and cultural diversity, and inherent contingency. Each is keenly aware of the evils of worldly life and becoming. But this recognition seems only to have steeled their scientific, philosophic, and moral resolve as the champions of worldly life. They call us to the way of earthly responsibility, however imperfectly understood. Moreover, they have unambiguously shown us the decided preference of paths of adventurous, enlightened ignorance over illiberal quests for certainty and final perfection.

APPENDIX

The final paragraphs on Whitehead's and Mayr's ethical convictions and calls to moral responsibility raise a nagging problem for many thinkers and philosophers, if not the rest of us. The essence of the problem, as I understand it, is that, if evolutionary, ecological nature is such an amoral, historically chancy affair, governed by a blind tinkerer rather than a cosmic designer, how can we arrive at any abiding and stable conceptions of the morally good (or evil) to define our moral responsibilities and guide our moral actions toward nature and our humanly cultural selves? (Aldo Leopold and *A Sand County Almanac* give us a stellar example of such an attempt.) Are we left to cultural conversations among our human selves, with no possible appeals to "objective" standards of the good or the bad that are independent of mere human, subjective preferences (the modern legacy of the original Cartesian split of body and mind)? In short, how can Darwinian naturalism (a form of critical realism) and a true ethics of moral responsibility go hand in hand and have a coherent connection with one another?

The problem cannot be treated with justice here. But I do think that we can make some brief headway. First of all, I think we must look at things from *within* Darwinian naturalism, not from within pre-Darwinian essentialist or culturalist ("idealist") perspectives, which defeat the enterprise from the start.

The chanciness and radical contingency of evolutionary and ecological process can be overplayed. Take the evolutionary two-step. Variation (genetic, somatic, or behavioral) may be random, but selection, natural and perhaps sexual, is a nonrandom process, even if carried out amid historical contingencies. Adaptational processes—adaptation to worldly environments—are not the mere work of chance. Moreover, amid much becoming and change, there are, as emphasized by Mayr, conservative or conserving factors at work in nature: historically engendered and more or less cohesive genomes; fundamental and enduring *bauplans* (body types or programs); ongoing adaptations tracking changing environments; the engendering of biological and cultural diversity, which, though not permanent, are certainly not ephemeral. In short, over evolutionary history and time, organic forms and capacities and ecological functions and processes emerge, with more or less robust staying power. Evolutionary, ecological, geologic time is not the hustle and bustle of everyday human life.

These forms, capacities, functions, and processes are no human creations. We can *recognize* that they have come into being without our creative interventions. Moreover, thanks to our having arisen from the same earthly history, we have decided stakes in this natural, historical drama. Given the natural and cultural capacities that have emerged in us through worldly history, we can recognize the achievements of historical "biocultural diversities" and, however imperfectly, their goodness, including their conduciveness to ongoing life and its flourishing.

If God could create the world in six days and see (declare) that his creation was good—he gave no rational reasons—we, differently situated in the historical, creative becoming of nature and culture, can analogously see (understand) that natural and cultural creation is good and ongoingly explore what is conducive to natural and cultural well-being. (Natural and cultural forms, capacities, and interconnections stun us aesthetically, spiritually, and morally, both positively and negatively.)

In sum, our moral worldviews, reflections, and judgments are rooted in, if not rationally derived from, the natural and cultural world. They are rooted in worldly realities (the gift of our being natural organisms living in the world) and are not merely aworldly, cultural conventions. Moral responsibilities to humans and nature are real and historically engendered "abiding interests of humankind," to borrow a phrase from Isaiah Berlin. Berlin wished to espouse moral pluralism (innumerable real ["objective"] values, such as freedom, justice, equality, and more) as against moral relativism (mere subjective preferences or tastes, with no essential reference to the real world). I think that moral naturalists—the Darwinians, the Mayrs, the Leopolds, if not also the Whiteheads—would concur. At the very least, they found in themselves the evolved capacities and need to judge the morally quick from the morally dead or pernicious.

The morally good and bad, right and wrong—natural and cultural—may in truth be a moving target, rather than an atemporal stasis, but its movements, especially its more abiding features, are real and relatively glacial. This must serve as the standard for the moral naturalist, in lieu of an eternal, essentialist form of the morally good. Whatever, what naturally and culturally *is* certainly *can be*. I am not sure that moral naturalism can give more reasons or more of an ultimate rational answer than this.

The proof is in the pudding. Are *A Sand County Almanac* and the reflections of Ernst Mayr, among many others, errant nonsense? Are they not persuasive, and do they not carry the ring of truth? Do they not relatively suffice, even while calling for further philosophic and moral explorations? Or do we still have to pine for the pre-Darwinian age of moral certainty and eternal absolutes? Or for more contemporary moral conversations among humans, forgetful of their bodily being and immersion in the world, modern heirs of idealist traditions, still trapped in Descartes' *res cogitans,* with no reference to a relatively objective natural and cultural good? (To be fair, there are modern democratic and moral deliberators who recognize their bodily embedded, worldly status. But there are all too many "old-timers," uncritically mesmerized by a theory of the social or cultural construction of reality.) For most (all) of us, the eternally objective good is beyond our ignorance or itself a chimera. The more extreme alternative cultural proposals strike me as Gogolesque: all talk and no reality (*Dead Souls*). Neither is the way of moral naturalists, who find, despite abiding ignorance, themselves and their imperfect world real and sufficient enough. We only need to realize that the good is housed in the home of becoming and change, rather than the realm of eternal being (Plato et al.). Or, rather, we should recognize with old Heraclitus that being and becoming are one and the same and that beauty, goodness, perhaps all values, are born out of worldly dynamism or strife. As the fifth-century B.C. philosopher said: "It is wise to know that all things are one, an Everliving Fire, kindling and extinguishing in measures." How about that for speculative, enlightened ignorance, declared at the very beginning of Western philosophy?

NOTES

1. The exposition of Whitehead's thought relies heavily on the classic essays "Nature Lifeless" and "Nature Alive" from pt. 3 ("Nature and Life") of his *Modes of Thought* (New York: Macmillan, 1938). The essays present the most accessible summary of Whitehead's mature critical and speculative thought. The reader could then consider *Science and the Modern World* (New York: Macmillan, 1925), *Adventures of Ideas* (New York: Macmillan, 1933), and then, finally, *Process and Reality* (Cambridge: Cambridge University Press, 1929). The latter is admittedly challenging but well worth the effort for those willing to stay the course.

2. Whitehead, *Modes of Thought,* 152. See also Whitehead, *Process and Reality,* 5–6.

3. Whitehead, *Modes of Thought,* 153ff.

4. Ibid., 130ff.

5. Ibid., 128ff.

6. Ibid., 132.

7. Ibid., 149.

8. Ibid., 132.

9. Ibid..

10. Ibid., 134–35, 154.

11. Ibid., 136ff.

12. Ibid., 136.

13. Ibid., 135.

14. Ibid., 148 ("Nature Alive").

15. Ibid., 138. See also Whitehead, *Science and the Modern World,* 51, 55.

16. Whitehead, *Modes of Thought,* 150ff.

17. Ibid., 151ff.

18. Ibid., 165.

19. Ibid., 151. See also Whitehead, *Process and Reality,* 25–26.

20. Whitehead, *Modes of Thought,* 159ff.

21. Ibid., 138.

22. Ernst Mayr, *One Long Argument: Charles Darwin and the Genesis of Modern Evolutionary Thought* (Cambridge, MA: Harvard University Press, 1991), 101ff.

23. Ibid., 12–47.

24. Ibid., 35–67; Ernst Mayr, *What Makes Biology Unique? Considerations on the Autonomy of a Scientific Discipline* (New York: Cambridge University Press, 2004), 26–28.

25. Mayr, *One Long Argument,* 53.

26. Ibid., 40.

27. Ibid., 40ff.

28. Mayr, *What Makes Biology Unique?* 24.

29. Ibid., 21–37 (chap. 3, "The Autonomy of Biology"), 67–80 (chap. 4, "Analysis or Reductionism"), and 159–68 (chap. 9, "Do Thomas Kuhn's Scientific Revolutions Take Place").

30. Whitehead, *Process and Reality,* 100–151.

31. Ibid., 36–39.

32. Ibid., 26–31.

33. Ibid., 36–39, 261–65, 403–13.

34. Ernst Mayr, *This Is Biology: The Science of the Living World* (Cambridge, MA: Harvard University Press, 1997), 16–23; Whitehead, *Process and Reality,* 76–151.

35. Mayr, *This Is Biology,* 268.

JOYFUL IGNORANCE AND THE CIVIC MIND

Bill Vitek

> There has never been a generation better educated than the one that
> ushered in the end of Athens.
>
> —Edith Hamilton

THE GIST

Advocating the virtues of ignorance is hard work. On the face of it, the proposition is preposterous to nearly everyone who hears it for the first time. People's response is that the claim must be a joke. It is not. Or that it's a spoof on the current political scene in the nation's capital, particularly in the White House. It is not that either. In the end, most folks become angry and say that there is already too much ignorance in the world and that it's making a mess of things. True enough.

What, then, is being praised, and how is it different from the ignorance with which we are most familiar and that we try earnestly to avoid? And how could ignorance possibly be a good trait, let alone a virtue or joyful? The short answer to this last question is that it depends on what ignorance is being called on to replace and why.

What is being replaced is a set of attitudes and beliefs—a worldview—about what can be known about the world, the methods by which this knowledge is acquired, of what this knowledge consists, who can possess this knowledge, and the moral boundaries governing the exercise of its use. Collectively, these attitudes and beliefs can be called a knowledge-based worldview (KBW) and can be seen operating most powerfully in the halls of science and engineering and with applications in agriculture, commerce, medicine, and even politics from time to time.

The reason for urging such a replacement is that these attitudes and beliefs about knowledge are increasingly being shown to be inadequate and dangerous. Praising ignorance, then, begins with a deep dissatisfaction with the KBW and moves slowly toward an understanding of what might productively and ethically replace it.

Proponents of this sort of ignorance—an ignorance-based worldview (IBW)—have rightly placed the bulk of the credit (and blame) for the KBW on the revolutionary and visionary work of two seventeenth-century thinkers: Francis Bacon and René Descartes. Between these two giants of pre-Enlightenment thought, I believe that it is Descartes' work with which there is most to be reckoned. It is not just his claim that human knowledge of the world is possible or his method of miniaturizing the world into discrete parts. Most importantly, and most dangerously—and, perhaps, unintentionally—Descartes likewise divides the community of knowers into ever smaller groups. For him, the pursuit of knowledge is first and foremost a solitary pursuit of a singular mind, a pursuit from which great power and good can come when it is correctly employed.

The influence of Descartes' work on the rise of scientific knowledge and its power and influence over other forms of interacting with the world are unquestionable. The Cartesian revolution marks the beginning of the individual as sovereign—first in science, then in economics and politics. Much good has come from this tripartite revolution of ideas, and we would be remiss to recommend a wholesale rejection. But so too has much trouble come, particularly in the misunderstanding of complex, living ecosystems, in the dangerous misapplication of partial knowledge in ways that are difficult to rescind or recall, and in the harmful effects of believing that the world is a laboratory or experimental playground.

My purpose here is to liberate the concept of ignorance in order to bring some sense to the idea that the solitary Cartesian mind is insufficient for the work ahead, not to mention a lonely place in which to linger. What replaces it is a civic mind focused on the pursuit of understanding, with others, in a living world.

WHAT WORDS MEAN AND HOW WE MEAN THEM

I'll start with some word work:

Science: Knowledge, to know, to discern.[1]
Know: To be assured of, recognize.

Knowledge: Assured belief, information, skill.
Recognize: To know again.
Assurance: To make sure, to secure.
Secure: Free from care, anxiety, safe, sure.
Ignore: Not to know, to disregard.
Ignoramus: Literally "we are ignorant" (formerly a law term). "Ig-
    noramus . . . is properly written on the bill of indictments by the
    grand enquest, empanelled on the inquisition of causes criminal
    and publick, when they dislike the evidence, as defective or too
    weak to make the presentment" (Blount 1970, s.v.).
Own: To possess.

These words and definitions tell a partial but interesting story. While we
cannot blame Descartes for the meaning of the words that we use in the
knowledge enterprise, both what he was after and what he was hoping
to avoid are nicely revealed in the words' etymological markers. "Science," be it natural, social, or political, is very much about knowledge,
and knowledge turns out to be about (1) recognition, a clear hat tipping
to Descartes' unquestioned assumption that the world and the human
mind are uniquely isomorphic; (2) assurance and, hence, an act of se-
curing (a form of control that strongly suggests a sense of ownership
and property); and (3) freedom from care and anxiety.

There seems likewise to be an obvious connection between *known*
and *owned.* Knowledge is something the owner has possession of,
sometimes exclusively, as in a discovery or a patent, sometimes in common with others. But the process of initial ownership, at least since
Descartes' time, has been solitary. The thinking mind isolates a natural
object or force, drives it into its smallest category or form, carefully
establishes boundaries around it, and sets to work to unlock its secrets.
The mind concentrates its forces, the object gives up its secrets, and the
knower becomes the owner of the secret.

The knower is an Al Haig—"I'm in charge here"—perceiver.[2]
And knowledge has a one-directional, observer-object, perceiver-per-
ceived quality. It's competitive too. The first one there is the propri-
etary owner of the newfound knowledge and receives the notice, the
prize, the patent, or the distinguished professorship. The process is
extractive as well. We see this most painfully even today in animal
experiments. Not all that long ago, vivisection was a common inves-
tigative and teaching tool, and "naturalists" identified, counted, and

extracted information about various species by killing and collecting individual members.[3]

Were we to search for an analogy, the pursuit of knowledge would seem very much like hunting. The best hunters study and stick to a small range, they stalk, they trick and deceive, they outwit their prey, the prey resists; the hunter, when successful, "takes" the prey in an act that can be pretty violent. When successful, the hunter feels victorious. The prey has given up its life (its secrets), which is now possessed by the hunter/investigator. What are Nobel prizes but the ultimate lion's head mounted over the fireplace in the library?[4]

Like hunting, the Cartesian method of acquiring knowledge may also be biologically predisposed in our historically/ecologically contingent, bicameral brains. Male brains may be especially inclined to a certain type of linear, boundaried, focused thinking that sees knowledge acquisition as a search for prey to be bagged (see, e.g., Baron-Cohen 2004).

The KBW has obvious power. The discoveries, cures, and inventions it has inspired have given its users some semblance of control, and occasionally dominance, over the powers and vicissitudes of natural forces. It has made possible a level of cultural capital (the arts, travel, and leisure) that, when not squandered on the mind-numbing leisure of television and video games, seems a positive benefit. That this worldview is also a benefit for a few, and is temporary, and is partial, and is potentially more dangerous than the dangers it overcame are all truths rarely mentioned. We should probably not vote to toss it out completely; rather, we should employ it as one tool among many, and not always the best tool for the job, and absolutely forbidden for some jobs, but certainly always and everywhere *only* a tool, not a bulwark for liberty or a free pass around limits.

We come now to the word *ignorance.* The task here is trickier since I want to shake loose the common usage and apply the term in a new way. The word *ignorance* gives us at least two lines of interpretation. First, as the definitions offered above state, ignorance is an absence of knowledge. But it's not just an absence. There's a moral tinge to ignorance, and it cuts a couple of ways. In the colloquial use of the term, an ignorant person is someone who doesn't know something but who could/should possess that knowledge had he applied himself. In this sense, ignorance is an easily correctable state for which the ignorant person is at least somewhat personally/morally responsible.[5] In the case of *ignoramus,* a second moral sense of ignorance exists. The "grand enquest" could refuse to present a case to the Inquisition because of weak or defective evidence. We have here an early and nonscientific application of the precautionary principle: when in doubt, do not proceed. Grand juries are impaneled every day in America for the same purpose. They consider evidence, vote on the strength of that evidence, and decide whether to halt proceedings or to return an indictment. Here, the quest for knowledge is not unlimited. If the evidence does not support the claims, they do not keep looking indefinitely but err on the side of freeing the potentially guilty rather than punishing the potentially innocent. We can see, albeit faintly, the important role of ethical restraint and personal character in a worldview where ignorance is a default position.

There is another way to go here as well, and it gets us closer, I think, to the ignorance we are looking for. The second meaning of *ignore* is "to disregard." Here, the willful state of ignorance is more rebellious. It refuses to look at the obvious. It ignores what is right in front of it. And it's glad to be doing so.

This conception is very nearly what we're after. Joyful ignorance is the state or activity of choosing to ignore the obvious, of ignoring what can easily or myopically be known. It begins as a willful, defiant, rebellious ignorance in the face of the obvious, of certainty, of security and control, of domination. "Can't you see it right in front of you?" asks the advocate of a KBW? Replies the advocate of an IBW: "I'm not looking right in front of me." Joyful ignorance is a first step toward a methodology for ignoring that obvious, loudest, and smallest piece of the universe that resembles the overachieving student whose hand always shoots up while yelling "Ster, Ster!"[6] Like that noisy, smart-in-a-certain-kind-of-way, but certainly overbearing student, certain parts of the world are often right in our face or too loud, too flashy, or too sexy to ignore easily.

An IBW is the claim that such willful ignorance may actually produce better skills, relationships, decisions, theories, policies, and long-term results than a worldview based on knowledge. How does it do this?

By refusing to focus on the smallest unit and, instead, striving to keep the perspective wide, practitioners of ignorance admit that they cannot know or have it all.[7] Ignorance therefore invites, indeed requires, a civic mind, a community of learners, and a time frame that stretches across, and connects, generations. The sole investigator has no place here. Ignorance requires us to work together, to communicate frequently, to share observations, and to fail often. Communities, social skills, and social capital are in strong demand here. More important, joyful ignorance elevates the status of the heretofore "object" of knowledge—that Cartesian chip of nature—to a fuller, more whole, contextualized, in many cases living, and in some cases conscious "Other." The language is tricky here since we may not want to call everything a subject or to put moral constraints on our work with minerals or water. Perhaps *relationship* is what we're after. The civic-minded practitioner of ignorance sees the world relationally, regardless of the matter studied. The IBW geologist, for example, sees her work with soils and rocks as a relationship, not only with them, but also with the wider realm in which soils and rocks exist and interact.

Sooner or later the geologist will bump up against the kinds of rules and virtues that govern relationships: questions of right and wrong, propriety, balance, foresight, faith, trust, and humility, to name a few. This is not to say that knowledge-based practitioners (e.g., scientists and engineers) are not governed by codes of ethics. They are. But the ethical codes loiter in the hallway waiting for a chance to speak up over the din of data points and findings. And, because the pursuits of science and technology are often so pinpointedly focused, it is understandably difficult for the researcher who studies the migration of oil globules through soil, for example, to see how ethics affects her work, other than the prohibition against plagiarism or "shoddy" science. In a worldview where knowledge is sufficient, ethical considerations are almost never directly on the playing field, rarely on the sidelines, and most frequently in the stands. The ethicists just slow down the game. An IBW, on the other hand, invites the other virtues into the game as equal partners. Slowing down the game is part of the game.

Finally, ownership (the *my* of *my patent* or *my property*) is replaced by fellowship (the *my* of *my family* or *my community*).[8]

If knowledge is a tool, ignorance is a perspective. The knowledge tool is used too often, too confidently, too absentmindedly, and sometimes quite destructively.[9] The ignorance perspective is too often sorely absent. It could provide us with plenty of information, understanding, wisdom, and happiness, in short, many of the qualities promised by the tool but undelivered in the main, or damaged along the way.

I compared knowledge to hunting and stalking above. An IBW, I think, is more like gathering. Gathering is a collective enterprise with larger ranges and boundaries. The "catch" is typically the fruits or eggs, not the plants or animals themselves, and, therefore, less violent. It is a multitasking activity, whereas hunting is usually about one prey on one day or during one season. Gathering is not particularly competitive, nor does it fuel collecting for its own sake or the trophy mentality so identified with hunting. Gathering/ignorance is less flashy but no less—and, arguably, more—productive in the long term. And gathering requires a deep understanding of terrain, seasons, and safe and dangerous foods and an awareness of the sort required by hunters. While I can't prove it, the gatherer, like the "ignorer," has a better sense of the larger terrain's influence on what is gathered. Fruits are grown from soil, rain, and sunlight; they grow well or poorly among other plants; they should not be picked too often or during certain times (restraint). Fruits are a product of an entire system's fertility, stability, and diversity. So, too, is the antelope. But the act of hunting, I think, is less able to state this lesson clearly.

Finally, if men are primarily hunters (and the creators of atomic weaponry, napalm, long-range rockets, etc.), then women tend to garden, gather, nurture, multitask, and more readily accept uncertainty and ambiguity. As science practitioners, women have done much better at seeing interconnections in nature, at letting nature "express" itself (e.g., Barbara McClintock's work),[10] and with the precautionary aspects of an IBW.[11]

These paragraphs represent a first pass. We'll need to do a lot more in terms of naming and describing an IBW. We'll also need to put it to use, promote it, and teach it.[12] I don't imagine that it will do anything magical for us, and I do imagine that it will encounter problems of its own. We should see it, not as a direct competitor with or an alternative to knowledge, and not as a tool to get actual work done, but, rather, as a perspective—at least at first—that we bring to our work. An IBW functions much like the perspective that my friend Clark Decker, an Ivy League–educated, former bank executive, fifth-generation dairy farmer, brings to his daily life. Clark somehow manages to laugh, to maintain soil fertility, milk production, and good relations with his neighbors, to love and honor his family, and to keep multitasking away in the midst of record-low milk prices, record-high energy prices, and a recent microburst thunderstorm that took out a third of his sugar maples. That is, by balancing his substantial knowledge and experience with the similarly substantial vagaries of weather, prices, and soil conditions, he keeps paramount the broader concerns of family, husbandry, and neighborliness while the world of bankers, economists, and surly weather patterns tempts him away from his core values. Such a perspective exists in many of us and still remains a possible alternative for that cultural artifact called the Western mind. Fortunately, we have in Aldo Leopold's life and work an exemplar of an ignorance perspective guiding the search for knowledge, understanding, and wisdom in the twin pursuits Leopold set for himself: "the relation of people to each other, and the relation of people to land" (Meine 1988, 51).

ALDO LEOPOLD'S CIVIC MIND

We have a pretty good sense of what a civic center or a civic initiative is, but probably we're less clear about civic education or civic agriculture or a civic economics or a civic mind. I'm not sure myself what these traditional terms would look like in their civic frameworks, but we can begin to imagine them. *Civic* means "belonging or pertaining to the citizen" and *mind* means "the understanding, intellect, memory." We can then ask: What would a theory/policy/institution/practice look like, or how would it be described or implemented, if it was considered either as necessary for citizens or as belonging to them and their shared pursuits?

A civic way of knowing requires a social network and setting. It requires others: as teachers in a tradition of learning; as fellow investigators within and across disciplinary boundaries; and as relational subjects of study. Like any social network, a civic mind is bounded by values and constraints that are derived not from individual members but, rather, from a shared understanding of the community. This communal aspect of a civic knowing limits the misplaced self-confidence and competition of the individuated mind in solitary pursuit of knowledge and strengthens the faith in common pursuits commonly undertaken and ethically applied.

Civic values and constraints are firm but generally not immovable or absolute. They, like the subjects under investigation and the stories and theories we construct about them, are likely to change as more information is gathered, more observations are processed, and more layers of complexity and interdependence are introduced. Aldo Leopold's ethical move from "predator as varmint" to "predator as interdependent and morally considered community member" is an example of movable and expansive value shifting.[13]

A civic mind pursues methods and skills for understanding and wisdom that can be made available across the community and not solely or exclusively for only certain members of the community. A civic mind ties together the sense that citizens are students of knowledge, including its limits, and that students of knowledge are citizens (members of interdependent, cross-cultural/species communities). Finally, because the civic-minded person is herself a citizen of the community that she likewise studies, she keeps the investigatory lens wide, the disciplinary boundaries fluid, the exits well marked, the larger implications well in mind, and the collaborator welcome mat in plain view.

It is Aldo Leopold's work and manner of thinking that supply most of the horsepower for this perspective. I can recommend no better example of a thinker whose understanding of the interactions of humans and nature was fueled by a joyful ignorance and wide-angle boundaries. Many readers are already familiar with Leopold's work. Those who are not will find in Leopold an exemplar of the ignorance-based perspective being recommended here. Some examples below will have to suffice:

- Leopold had a voracious intellectual appetite, and he read widely across disciplines. He brought philosophy, literature, history, and biblical scripture to his scientific endeavors.
- He did more, I believe, than any other twentieth-century thinker in reviving *citizen* and *citizenship* as terms that cut across species, in bringing the misadventures of humans in the natural world into sharper focus and in calling for educational, ethical, and political systems focused on humans as social, not solitary, beings.
- He resisted "either-ors." He encouraged the farmer and the hunter to work together; he spoke to diverse groups—farmers, civic groups, engineers, and college students—about what they could do in the name of environmental conservation; and he sought to reconcile conflicting environmental policies (e.g., the conflicts between what he called "wild lifers" and game farmers).
- He was willing to revise, reject, and reconsider his earlier conclusions. His reconsidered views on predators and on the role of fire in forest ecosystems are two obvious examples.
- He saw the need to radically revise science education with two central goals in mind: to better educate citizens about the world

they live in and to broadly educate scientists beyond their disciplinary teacups.

- He loved the natural world he studied and wasn't afraid to say so: "I love all trees, but I am in love with pines"; "Our geese are home again"; "We mourned the loss of the old tree"; and "Now I am free to grieve with and for the lone honkers" (Leopold 1966, 74, 21, 9, 22).
- He seemed to delight in his ignorance about the natural world and wished not to know it all. Knowledge, ignorance, faith, love, and understanding play equal roles in his work: "If I could understand the thunderous debates that precede and follow these daily excursions to corn, I might soon learn the reason for the prairie-bias. But I cannot, and I am well content that it should remain a mystery. What a dull world if we knew all about geese" (Leopold 1966, 22).
- His ethical, ecological, and aesthetic boundaries were intentionally wide, and they discouraged the strong temptations of small discoveries made possible by the purely scientific mind. Yes, Leopold was after mechanisms, but they were land mechanisms and ethical mechanisms, the big ones: "Only those who know most about the complexity of the land organism can appreciate how little is known about it" (Leopold 1966, 190).
- Though rarely dire or shrill, Leopold urged caution on almost every page. Most famously: "To keep every cog and wheel is the first precaution of intelligent tinkering" (Leopold 1966, 190).
- He credited his nonhuman teachers often, whether they were his dogs, a tall grass prairie, a buck he missed, or the small fish he caught: "We are not scientists. We disqualify ourselves at the outset by professing loyalty to and affection for a thing: wildlife. A scientist in the old sense may have no loyalties except to abstractions, no affections except for his own kind" (quoted in Meine 1988, 274).
- He hunted and fished, but he loved and respected his prey, practiced strong ethical constraints even when the law required none, and saw his work as benefiting not just humans but the land as a whole: "My dog was good at treeing partridge, and to forego a sure shot in the tree in favor of a hopeless one at the fleeing bird was my first exercise in ethical codes. Compared with a treed partridge, the devil and his seven kingdoms was a mild temptation" (Leopold 1966, 129).

This summary should block claims like "The civic mind is power-less" or "If ignorance is our strong suit, then what?" The civic mind is certainly not powerless. Leopold's scientific and amateur observa-tions contributed to the ecological encyclopedia. He can be credited with creating a new field of study (wildlife management), with advanc-ing the field of ecology, and with suggesting new directions in ethics, economics, and education. Most important, Leopold never forgot for long about the wider, infinite world in which he was living and work-ing. He delighted in directing the reader's attention to the Pleiades and the Pleistocene, distant places and times to which each of us is still and essentially connected. "Dust to dust, stone age to stone age," he says in his essay "Song of the Gavilin." He continues: "It was appropriate that I missed [the buck], for when a great oak grows in what is now my gar-den, I hope there will be bucks to bed in its fallen leaves, and hunters to stalk, and miss, and wonder who built the garden wall" (Leopold 1966, 160). Leopold's long view is slow, wide-angled, precautionary, com-munal, ethical, reverent, and joyful, full of partially understood details, and lacking irrefutable answers. His work as a philosopher is as good as it gets in the twentieth century, or any century, if we dare step out-side the usual confines of the philosophical canon. His work is close at hand, accessible, and respectable. We don't have to start from scratch. As an example of what I think we're looking for, Aldo Leopold's work turns out to be an exemplar, a model that can be usefully studied and replicated.

A CIVIC EDUCATION "TO DO" LIST

Aldo Leopold understood that civic minds do not spring forth fully formed and that they do not get much developmental help in our current educational system. In a note to his students he wrote: "The objective is to teach the student to see the land, to understand what he sees, and enjoy what he understands. . . . Once you learn to read the land, I have no fear of what you will do to it, or with it. And I know many pleasant things it will do to you" (Leopold 1991, 301, 337). The work required to move an IBW into the classroom, and to begin training a new genera-tion of civic minds, requires large shifts and tremors in what we now consider a good education.

Fortunately, there is some good news here. Educational models al-ready exist that introduce students to big, chunky questions that lack simple, well-defined answers. One of the best is problem-based learning.[14] Citizen science, first encouraged and promoted by Leopold, is also sufficiently well established.[15] New interdisciplines with sufficiently broad and ignorance-inviting perspectives are being developed—complete with journals, professional societies, and curricula. They include ecological economics and industrial ecology.[16] And university faculty members are increasingly encouraged to leave their silos to work with community members on projects that require real communication across disciplines and a big-picture perspective.[17]

But there's much more to be done. Here's a short list:

- We need more exemplars, both historical and contemporary, of civic knowledge and an IBW in order to demonstrate their viability and reasonableness.
- We need to keep philosophers of the right sort around and invite them in.[18] Philosophy is ideally the best discipline for big-picture thinking and for the practice of healthy skepticism, and it is the best tonic for academic smugness. I've heard Wes Jackson say that philosophy will be the most important subject in the coming decades. I'll risk the charge of self-server and say that I think he's right.
- We need to keep and expand liberal arts education. This is treasonous language in the money-strapped academy and anxiety-ridden suburbs these days, but, without the arts, history, literature, and philosophy, we have no hope of getting this new generation ready for the IBW switch to come.
- Science and engineering students need more courses in their respective histories, including failures, mistakes, ruses, false paths, ridiculous conjectures, and unseemly partnerships—in short, a full and thoughtful dose of the limits and ample ignorance in their chosen fields.
- We need to retool graduate education. This is the weakest link in the transition. Young people can go from kindergarten through college with all manner of projects, teamwork, and interdisciplinary messages, only to find themselves in a graduate program devoted to the single pursuit of the smallest unknown unit and receiving disappointed looks when requesting courses outside the department. There are a few exceptions out there, but they're not tide turning.[19] We should require that every Ph.D. student know

the history of her field, appreciate the larger impacts of her research, and be able to deliver a lecture to an educated, nonacademic audience.

- We need to break down the disciplinary and professional barriers in our colleges and universities and in our professional societies. We need more faculty lounges and cafeterias, more general lectures instead of departmental seminars, more "get-real" problem-solving sessions at our annual conferences, and more cross-conference fertilization.

- Those of us in the academy with the ability to do so (i.e., tenured professors and fearless administrators) need to work to change the reward structures and to seek broader definitions of *research, publication,* and *service* that extend beyond "specialty," "peer review," and "campus committees."

- It wouldn't hurt to find a bit more joy in our work as well. Environmentalists generally are considered a pretty "down-in-the-mouth" group, and, in trying to get others to take the issues seriously, they tend to bring the rest of us down too. And the success rate of this approach is hardly impressive. Aldo Leopold was clearly concerned about the plight of Western civilization, and his writings made this clear. But he likewise found and described the joy he felt in nature. Let's promote the excitement and enthusiasm we feel about the paradigm-switching work ahead. Let's excite and challenge a new generation of thinkers to dive right in.

- Over time we will need to develop the standards of evidence and argument necessary for the new worldview. Quantitative and qualitative measures for ignorance and the civic mind will help advance our work. We cannot be just about undoing a Knowledge-Based World View. Our movement toward an Ignorance-Based World View requires standards, metrics, terms, concepts, definitions, and a methodology. There's plenty of work to keep a generation or two of graduate students busy.

- Like any paradigm alternative, ours will need to work reasonably well to replace what came before it and to both explain what a KBW explains and additionally explain what it fails to explain (and to avoid the problems caused by a KBW). It will need to do more, not less, and to do it better. That's how a new paradigm replaces an old paradigm.

- It would be prudent to begin finding some historical case studies that demonstrate how a civic-minded IBW would have solved a problem differently or avoided an error (e.g., pesticides, nuclear power, plastics, leaded gasoline). These cases should likewise include current debates regarding genetics, climate change, nuclear waste disposal alternatives, male infertility, etc.
- I think we should also hold the line on systems thinking and resist seeing it as anything more than a KBW on a larger, computer-modeled scale.[20] There are noble intentions behind the assumptions of systems thinking, but too often it replaces linear thinking about parts with computer-modeled holistic thinking about more parts. The abstractions added by the models tend to leach out and render suspect whatever holistic conclusions are reached.
- We'll need to find the proper role for a knowledge-based view. Certainly not a *world*view, but a role appropriate to the power (P) of knowledge (K), divided by its limitations (L). Something like $K = P/L$, where we don't proceed, or we proceed only slowly, when $K < 1$.

And, while we're at it, we should start celebrating the new ignorance-based worldview by giving out substantial prizes to its most influential proponents. There's nothing like a fancy prize, some cash and prestige, some press, and some envy to goose the new paradigm. People will want to know what all the fuss is about.

CODA

We are creatures of evolution, and the civic mind is a work in progress. It is shaped by physical and cultural forces that exceed our individual attempts to control or comprehend them. Yet human consciousness can glimpse these forces at work in the world and occasionally glean patterns, make predictions, and improve local conditions. Up against our small insights and successes is a vast and vigorous unknowing, constantly on the move in patterns and cycles that take us out beyond galaxies and our very conceptions of galaxies. In such places and against such forces there is little to be done well to change things, let alone to improve them substantially. Joyful ignorance is a wide-angled and lighthearted recognition that, against such odds, the best strategy is to join with others in the slow and always incomplete pursuit of understanding and to enjoy the ride. Our collective work can do little to make big changes, but, over the long haul, and working and living with others, it can contribute to building "receptivity into the still unlovely human mind" (Leopold 1966, 295). That's plenty good enough.

NOTES

1. All definitions employed in this essay are taken from Skeat (1974).

2. Moments after the March 1981 assassination attempt on President Ronald Reagan, Al Haig, then secretary of state, wrongly concluded and then announced: "I'm in charge here."

3. There's the story—perhaps apocryphal—about the graduate student who killed one of the world's oldest trees by trying to measure its age.

4. Here's another interesting parallel between hunting and knowing, and anecdotal to be sure. Many young boys begin their hunting careers blasting away at nearly anything that moves. Over time, some of these boys turn into men with a deep respect for, and an ethical relationship with, the prey they hunt and will forgo a shot for reasons that have nothing to do with range, size of the prey, and so on. They give up the hunt per se for something nobler. Similarly, the winners of the Nobel Prize in Chemistry who visit our campus annually as part of a distinguished speaker series give public lectures extolling the virtues of, and the necessity for, a broader, interdisciplinary approach to science. Having won the most prestigious prize for successful Cartesian thinking, they now see the need for another, broader approach. I am not so certain that the scientists in the audience who have not won such a prize agree.

5. We would not call someone with a learning disability ignorant.

6. *Ster* is short for *Sister* and was used by many of my classmates in the Catholic school I attended. I'm sure the word is well-known among Catholic school alumni across the nation, especially those of us who were not so fast with the answers.

7. This wide perspective and the admission of ignorance set our ignoramus apart from those who willfully refuse to look at the bigger picture or refuse to consider new and/or contrary evidence precisely because they think they know it all already. "Good" ignorance willfully ignores the obvious in favor of something larger and broader. "Bad" ignorance willfully refuses the larger and broader in favor of a more narrow perspective.

8. It is fellowship, not ownership, behind Leopold's exclamation in his "March" essay: "Our geese are home again!" Fellowship is also behind his remark in "July": "Books or no books, it is a fact, patent both to my dog, and myself, that at daybreak I am the sole owner of all the acres I can walk over" (Leopold 1966, 21, 44).

9. This does not imply that a knowledge-based methodology has no place in the cultural toolbox. I think it does. What it can't do is serve as a worldview or usefully (or safely) advance all manner of pursuits. We'll recognize its downgrading from a worldview to a methodological tool when we see fewer natural scientists, economists, and social scientists hogging the news hour with their findings.

10. "I know [my corn plants] intimately, and I find it a great pleasure to know them." Or: "It might seem unfair to reward a person for having so much pleasure over the years, asking the maize plant to solve specific problems and then watching its responses." See http://www.brainyquote.com/quotes/authors/b/barbara_mcclintock.html.

11. Rachel Carson, Sandra Steingraber, and Devra Davis are obvious examples.

12. The KBW is a bigger berg than even we ignorance proponents think, and our academic hot air, by itself, won't reduce it by much.

13. It's dicey to talk about shifting values, but here's a rule of thumb: we should be more inclined to accept values that change as a result of increasingly inclusive processes and methodologies and less inclined to accept values that change as a result of increasingly exclusive processes and methodologies (gay marriage vs. the Patriot Act, e.g.).

14. For an example, see Rhem (1998).

15. For an overview of the concept of citizen science, see http://www.birds .cornell.edu/LabPrograms/CitSci.

16. For examples, see http://www.umich.edu/~nppcpub/resources/compendia/ ind.ecol.html; and http://www.ecologicaleconomics.org/about/intro.htm.

17. For examples, see http://www.psc.cornell.edu; and http://www.clemson .edu/public.

18. The "right" philosophers are those without too much academic, professional, or thematic baggage. Those best suited for the job may have had little or no formal academic training in philosophy.

19. Dr. Tom Theis, the most visionary engineer I know (and I know many engineers), was required to take philosophy seminars in both his undergraduate and his graduate engineering programs at the University of Notre Dame. He created a new Ph.D. program in environmental manufacturing management at Clarkson University that required courses across the disciplines and a multiple mentoring committee instead of the traditional faculty adviser. He now directs the Institute for Environmental Science and Policy at the University of Illinois at Chicago (http://www.uic.edu/depts/ovcr/iesp/index.htm). I'm sure there are other examples out there, but we need more of them.

20. On systems thinking, see Aronson (1996–1998).

REFERENCES

Aronson, Daniel. 1996–1998. "Introduction to Systems Thinking." http://www. thinking.net/Systems_Thinking/Intro_to_ST/intro_to_st.html.

Baron-Cohen, Simon. 2004. *The Essential Difference: The Truth about the Male and Female Brain.* New York: Basic.

Blount, Thomas. 1970. *Nomo-Lexikon: A Law Dictionary and Glossary.* Los Angeles: Sherwin & Freutel. Originally published in 1670.

Leopold, Aldo. 1966. *A Sand County Almanac.* New York: Ballantine.

———. 1991. *The River of the Mother of God and Other Essays.* Edited by Susan L. Flader and J. Baird Callicott. Madison: University of Wisconsin Press.

———. 1999. *The Essential Aldo Leopold: Quotations and Commentaries.* Edited by Curt Meine and Richard Knight. Madison: University of Wisconsin Press.

Meine, Curt. 1988. *Aldo Leopold: His Life and Work.* Madison: University of Wisconsin Press.

Rhem, James. 1998. "Problem-Based Learning: An Introduction." *National Teaching and Learning Forum* 8, no. 1 (December), http://www.ntlf.com/html/pi/9812/pbl_1.htm.

Skeat, Walter W. 1974. *An Etymological Dictionary of the English Language.* Oxford: Clarendon. Original work published 1879–1882.

PART FOUR

Applications

I Don't Know!

Robert Root-Bernstein

I still have the copy of Gerald Ames and Rose Wyler's *The Giant Golden Book of Biology* that I was given when I was eight years old. Two characteristics have made it dear to my heart. The illustrations by Charles Harper, highly abstract and stylized yet illuminating, appeal to my visual sensibility. Equally compelling is the concluding paragraph of the foreword by George Wald. "I knew I would like this book," he wrote, "when I read on the first page: 'Questions are just as important as answers.' Science is a way of asking more and more meaningful questions. The answers are important mainly in leading us to new questions. So try to learn some answers, because they are useful and interesting, but don't forget that it isn't answers that make a scientist, it's questions."[1]

George Wald was listed at the end of the foreword as being a "Professor of Biology, Harvard University." If I hadn't known before, I certainly learned from the book that biology was the science of life. I'm not sure that, at the age of eight, I knew what a professor was. I'm sure I'd never heard of Harvard. But I can remember deciding at the age of fourteen, when Wald was awarded a Nobel Prize, that he probably knew what he was talking about. That's also about the age I knew that I would become a biologist.

I'm not sure how much to blame Ames, Wyler, and Wald for the impish interrogator that I have become. After all, we have a tradition in our family called "doing a Bernstein" that involves getting up in public and questioning the dearly held assumptions of our various professions. Questioning was something that happened every evening at dinner, when my father would regularly discount everything we took for granted. He was a skeptic. It was good training for learning to think for oneself.

My earliest memory of "doing a Bernstein" occurred, coincidentally, the same year Wald got his Nobel Prize, when I was fourteen. It involved geometry, not biology, however.

The problem must have arisen during the first week or so of geometry class. The teacher defined a point as a dimensionless object with no size, infinitely small. Lines, she then explained, are made of an infinite number of points, planes of an infinite number of lines, and solid objects of an infinite number of planes. Standard textbook stuff, definitions, in fact, not open to question.

Yet I objected. Having read Norton Juster's *Phantom Tollbooth* many times, I knew full well that neither his hero, Milo, nor anyone else could ever scale the "Infinite Staircase," for there was always one more step, and one more, and one more, and one more after that . . . By analogy, no matter how many infinitely small points I placed between two others, there would always be room for another, and another between them, and another between them, and another between them . . . So, no matter how many points I accumulated, I wouldn't be able to make a continuous line. In fact, the more points I added, the more gaps there would be, sort of like Xeno's paradox: move halfway to your destination, then half the remaining distance, then half of that, then half of that, and, of course, you never get there. So, if I could never fill in the gaps between two points, I couldn't make a line; no lines, no planes; no planes, no solids; and you can see where that led!

Apparently, I argued with the teacher for much, if not all, of a class period, and I can still remember my dismay at being told that discussion was at an end; I would have to accept the definitions.

I never did. I learned to use them, but I never accepted them. I continued to question every definition and axiom. I quibbled and quarreled. I ended up with a B in the class, the only B I got in junior high school. I even flunked a quiz that term, something I had never done before and never did again. And yet, and yet, at the end of the year, despite my relatively poor performance, despite several friends who had aced that and all their other math classes, I was awarded the school's mathematics prize!

It would be decades before I understood what I was grappling with at the time. If the geometry teacher knew, she didn't tell me. And my father, who was a math major in college, never let on. Perhaps they didn't know, either. But they let me question. *They let me question!*

And what I slowly learned was that I had been asking the kinds of questions that had led to mathematical revolutions. What if parallel lines *do* meet at infinity? This question led to the development of the non-Euclidean geometries that were so important to Einstein's discoveries. What if there are different *kinds* of infinity? This question led Cantor to rethink the entire field of sets. You can also assume, as I wanted to do, that points are not infinitely small and that lines cannot be infinitely divided. It's just that what you get isn't geometry or calculus any more. It's a different kind of mathematics called net theory, a kind of math most of us are never exposed to during our entire lives. My father would introduce me to net theory in college, and I would become one of the first to apply it to modeling biological systems,[2] prepared by a decade of iconoclastic questioning.

The next memorable "Bernstein" that I pulled occurred in 1973, when I was a junior in college. It was unintentional. I was sitting in a course called "Principles of Biochemistry," learning about the structure of DNA. We were introduced to the standard textbook material about DNA being a double helix held together by complementary base pairs, and the professor put up lovely drawings of this double helix unwinding into two separate strands, each of which formed a template for the replication of a new double helix. This was what James Watson and Francis Crick had called "the secret to life" when they discovered the double helical structure in 1953. It certainly looked impressive on the screen in the lecture room and equally compelling in the illustration in the textbook.

But the mention of unwinding bothered me. I'm a hands-on type of guy. I play with things. One of the things I'd played with as a kid was unwinding a rope to see how it was made. I had discovered that unwinding a rope is no easy task. It takes energy. It takes patience. And it takes great care. The free strands of a rope or a string tend to get raveled and knotted as the cord untwists (for every untwist, there is an equal but opposite twist!). I mentally applied my experience to the problem of unwinding DNA.

To give you some idea of the problems involved in unraveling DNA, if one were to model a single strand of DNA in a single chromosome as a rope, it would stretch from the earth to the moon. This million-mile-long rope is all curled up inside the nucleus of a cell the size of a house and needs to be untwisted within this incredibly cramped space into two free strands, neither of which can get tangled with itself or with the other stand. I raised my hand, laid out my case, and asked the professor what kind of a mechanism could possibly carry out such a procedure.

After all, there's no little boy inside the nucleus to do the untwisting or to keep the two strands untangled.

To say that I was surprised by the response would be an understatement. The professor turned red, spluttered for a moment, and then, in front of about two hundred other students, screamed at me: "*Just take it for granted that it works!*"

I knew right then that I had struck a nerve. A bit of investigation revealed that the professor himself was working on this very problem and had not yet figured it out. Neither, it turned out, had anyone else at the time. Why the professor couldn't just have said so, I don't know. Perhaps he was worried, as Watson and Crick were in some of their early papers, that the DNA double helix couldn't be unwound and that the whole model would have to scrapped. Perhaps he was as uncomfortable as I still am with the "solution" that eventually emerged, which suggests that the DNA is cut by enzymes into small fragments that can easily be unwound and that these fragments are then pasted back together into the separate strands. (Can nature really perform such an important function in such a piecemeal fashion?)

In any event, I learned three important lessons. The first one was never to trust a picture.[3] Every picture of DNA unwinding that I have ever seen, even to this day, misportrays the process: not one shows the cutting and pasting that current theory accepts. And that means that every student who accepts what they are seeing in these pictures is learning an incorrect model that glosses over the very real and important unwinding problem. The second lesson was how hard it is to see what isn't there but should be. The absence of the unwinding problem from discussions of DNA function has resulted in the fact that most scientists don't even know that the problem ever existed. It is not mentioned in Watson's famous autobiography, *The Double Helix;* it is not mentioned in Crick's autobiography, *What Mad Pursuit;* it is not mentioned in any of the standard histories of the discovery of the double helix by Robert Olby, Horace Judson, or Matt Ridley; it is not mentioned in any introductory modern biology or biochemistry textbook that I have seen.[4] I believe that no one really wants to face up to the problem, so we pretend that it doesn't exist. And the third lesson was that experts are people (like my professor) who are unable to say: "I don't know."

I don't trust experts. Whenever a problem persists for any significant length of time, that means that the experts not only don't have the solution but also have failed to ask the right question.

I was to learn just how important asking the right question can be in graduate school when one of my students in a biology lab asked me a question I couldn't answer.[5] I should immediately clarify the fact that I was, and still am, often asked questions that I can't answer and that, rather than screaming at my students, "Just take it for granted that it works!" I have a rather different response. I say: "I don't know! Let's figure out how we can find out." Unlike many of my colleagues, I have no need to be the ultimate authority on all things in my subject. In fact, I'm not much interested in what is already known. What attracts me to science is what we don't know, that thin layer of living questions that deposit what the physicist William Lawrence Bragg called the sturdy but dead coral of scientific knowledge. So my object in teaching is to encourage questioning and then to enlist the questioner in the living process of finding answers. This turns students into investigators and captures their imaginations.

Duncan Fisher was the extremely bright freshman at Princeton University who asked me the question that neither he nor I nor anyone else has yet answered. My recollection is that Mr. Fisher's parents both had M.D.-Ph.D.'s, that he had taken Advanced Placement Biology in high school, and that he had no business being in the lowest-level introductory biology course. But there he was, and it was a joy to have him since he brought a wonderful sense of humor and a plethora of biology- and medicine-related puns and jokes to the lab every week.

Mr. Fisher's question came in response to a section of the course on human blood typing that I had taught many times and thought I had mastered. I explained to the class the well-known fact that there are four basic blood types: A, B, AB, and O. I described how these blood types are inherited as variations in the proteins on red blood cells. I warned that, if you are given a transfusion with the wrong type of blood, your immune system will respond by trying to destroy the mismatched blood, resulting in coagulation of the cells in the circulatory system and death. Mr. Fisher immediately raised his hand and said: "I don't understand."

I was shocked. Mr. Fisher was my classroom litmus test. If Mr. Fisher hadn't understood, nobody in the class had understood! I immediately began reviewing in my mind what I had just said to see whether there was something wrong or something I had left out but found nothing obvious. So I encouraged Mr. Fisher to elaborate. He summarized without a flaw what I had just told the class, emphasizing the fact that each person's body can determine whether it has been exposed to an incompatible blood type. Then he added the kicker: "So how can this be consistent with the Central Dogma of molecular biology?"

If I had been shocked before, I was now speechless. What was the relationship of the Central Dogma of molecular biology to blood transfusions? I couldn't even imagine what Mr. Fisher's question meant! Risking the appearance of stupidity, I replied: "You'll have to explain."

"The Central Dogma," Mr. Fisher proceeded, "states that information in a biological system flows from DNA to RNA to proteins. Once it gets into a protein, it can't get out again. Right?" I agreed. DNA can make copies of itself, and so can RNA, and DNA can act as a template for RNA, and vice versa, but proteins can't make copies of other proteins, and proteins can't (as far as we know) be templates for making RNA or DNA. Once genetic information gets into a protein it doesn't flow back to RNA or DNA. Francis Crick, one of the discoverers of the DNA double helix, had named this irreversible process the "Central Dogma of molecular biology" because there was not (and never could be) any evidence for it. It was a simplifying assumption that, like a religious dogma, had to be accepted on faith. This was something we'd covered ten weeks earlier in the class. I still didn't see the connection to blood types.

"OK," Duncan continued, "but you've just told us that the ABO blood system is based on inherited differences in proteins on red blood cells. If information can't get out of proteins, then how can my immune system, or yours, tell the difference between matched and mismatched blood? Doesn't that discrimination require information to pass from the donor's proteins to your immune system, which has to check it somehow against your genetic code, thereby violating the Central Dogma?"

Wow!

This wasn't the kind of question I could help Mr. Fisher look up somewhere. There was no simple experiment that would suddenly yield enlightenment. This question linked two areas of biology that textbooks treated so independently that I wasn't even sure how to go about thinking about the question! I told him I'd get back to him in a week or two when I had a better grasp on the issues.

To make a long story short, I'm still working on defining the issues raised by Mr. Fisher's question. It involves questions like: "What is information?" "If information flows into a protein from its encoding DNA and RNA, then how is that information stored?" "Why should we not be able to get the information back out of a protein? Is it in some way lost?"

"Doesn't the immune system get information out of proteins when it differentiates between 'self' proteins and 'nonself' ones?" "How *does* the immune system differentiate between 'self' and 'nonself'?" "Why should we accept a dogma without evidence or experiment, even if it was enunciated by Francis Crick?" and "Doesn't putting a dogma at the center of molecular biology undermine its scientific basis?"

My first scientific publications were, in fact, challenges to the Central Dogma because I cannot see any rational reason to accept untestable assertions in science.[6] I proposed mechanisms by which information might, in fact, flow between proteins or from proteins to RNA or DNA. They became the foundations of a field of biology now known as "antisense proteins" that has several active investigators around the world.[7] I am fairly certain that these papers were crucial factors in earning me one of the first MacArthur Fellowships in 1981. This must have galled Crick. By the time the papers appeared in print (a long and nasty process in which I was pilloried by more than one scientific reviewer!), I inhabited the only office on a floor of the Salk Institute for Biological Studies that also housed none other than Crick himself. I saw or heard Crick daily for over three years. He spoke to me only one time, which was to tell me that he had no time to speak to me, ever. Such is the vanity of great men!

What I learned from this experience is that some questions are taboo. But, having been awarded a MacArthur Fellowship nominally "to do things that I would not otherwise be able to do," I naively decided that this type of questioning was exactly what I should do. And, since I was already studying the immune system, I did yet another "Bernstein," this time related to the AIDS epidemic. When the human immunodeficiency virus, HIV, was isolated and proclaimed the cause of this new disease, I decided to ask whether a simple virus could be the cause of such a complex syndrome. My motivation was not pure perversity (although I'm sure some of my colleagues thought so) but, rather, the unusual training I had gotten from Jonas Salk as a postdoctoral fellow in his laboratory.

Salk was the man who led the team that developed the first successful polio vaccine. When I worked with him, he had moved on to the study of multiple sclerosis, a disease with which he, unfortunately, had no success. But what we learned in the lab was very interesting with regard to modulating the immune system. The same kinds of things that can stimulate the immune system in small amounts can overwhelm and shut it down when delivered frequently or in very large doses. For example, even properly typed blood causes serious immune suppression when transfused into a patient. The more blood the patient receives, the more immunosuppressed the patient becomes. During the 1980s, surgeons at the University of California, San Francisco, had begun to use blood transfusions to help transplant patients accept their transplants more readily. So it was obvious (at least to me) that people exposed to HIV through blood transfusions were getting not one immunosuppressive agent (the virus) but at least two: the virus *and* the blood. I began to wonder whether other people at risk for AIDS were also exposed to immunosuppressive agents in addition to HIV.

They were (and are!).[8] Patients requiring blood transfusions almost always receive anesthetics and opiate drugs as part of their surgical treatments, and these agents are immunosuppressive. Most of the drugs that addicts use inhibit immune function. Many of the chronic infections associated with reusing unclean needles impair the immune system. So does the malnutrition that often accompanies drug abuse and that occurs endemically in the impoverished areas of the world where AIDS is rampant. Tuberculosis, malaria, and sexually transmitted diseases such as hepatitis decrease immunological activity, and, the more such diseases a person contracts, the more impaired immune function becomes. The impure blood-clotting agents used by hemophiliacs during the 1970s and 1980s were also found to be immunosuppressive and often contained multiple viruses (such as hepatitis) besides HIV. And the oldest experimentally defined cause of immune suppression is semen, the primary carrier of HIV. As early as the 1890s, it was demonstrated that, if semen enters the blood stream, the immune system is compromised. Subsequent research showed that immunological exposure to semen is very common in unprotected receptive anal intercourse (in both men and women) but, for a variety of anatomical and physiological reasons, very rare in either oral or vaginal intercourse.

What I had found, in sum, was that the people most likely to contract HIV and then to develop AIDS were people who were already at far greater risk for immune suppression than is the typical human being. I therefore asked myself: "What if HIV is not the sole cause of AIDS but, rather, takes advantage of, or works synergistically with, these other causes of immune suppression?" On the one hand, such a possibility makes AIDS much more difficult to prevent and treat since the cause would no longer be a simple virus acting on its own. On the other hand, if HIV is most likely to take root among people with existing forms of immune suppression, then cleaning up blood and blood products (as has already been done!), preventing or treating malnutrition, drug abuse, hepatitis, malaria, tuberculosis, and sexually transmitted diseases, and using safer sex practices to prevent exposure to *semen* might have a very significant impact on the occurrence of AIDS. In fact, I have argued that it might be less expensive and of more benefit to the health of more people around the world to target the underlying causes of immune suppression than to target HIV.

Be that as it may, the scientific question is clear: How do we know that HIV is both necessary and sufficient to cause AIDS if every person who contracts HIV and goes on to develop AIDS has a multitude of other immunosuppressive factors acting on them? The answer to this question is, unfortunately, still elusive. Thus far, biomedical scientists have, with only a few exceptions, refused to design controlled experiments or studies that will yield answers. A notable exception is Luc Montagnier. Montagnier is the codiscoverer, along with Robert Gallo, of HIV. He has maintained since 1990 that HIV is not sufficient, on its own, to cause AIDS. He has argued, as I have argued, that HIV requires what he calls "cofactors" to stimulate its activity.[9] What neither of us understands is why the failure to halt the AIDS epidemic or to cure people with AIDS has not led more people to ask, as we have asked: "Do we really know the things about AIDS that we think we know?"

My questioning of AIDS dogma created many problems for me professionally, but it also created one wonderful opportunity. It brought me to the attention of Drs. Marlys and Charles Witte, a pair of surgeon-scientists who were also questioning many aspects of the HIV-equals-AIDS dogma, and it prepared me for the Curriculum on Medical Ignorance that they were devising in collaboration with the philosopher Ann Kerwin.[10] I was immediately fascinated by their classification of types of ignorance: things we know we don't know (overt ignorance); things we don't know we don't know (hidden ignorance); things we think we know but we don't (misknowns); things we don't think we know but we do (hidden knowledge); questions we aren't allowed to ask (taboo questions); questions that provoke answers we weren't allowed to consider (unacceptable answers). Boy, was I ever familiar with most of these, especially taboo questions! The question I had, however, was the one we all recognize in science as being the ultimate compliment to innovators: Why hadn't I been clever enough to think up such a taxonomy myself?

I did the next best thing. I started to help the Wittes think about how to use their taxonomy to train better questioners. I have focused my efforts on the sciences[11] but see no fundamental reason that the same approaches should not work in the social sciences, the humanities, and even the arts. Everyone, after all, deals with problems, and only a problem properly posed as an answerable question has any hope of solution.

I have had two types of useful insights. One is to realize that there are at least two additional types of ignorance that are of particular importance to scientists. One I call "unwanted visitor" problems. For scientists, nature is a huge puzzle that we can deal with only by breaking it down into bits and pieces. One general problem every scientist faces is which bits and pieces belong to which small pieces of the puzzle. There is no guide that tells us infallibly which pieces should go together, and very often we end up trying to put together pieces that don't fit. So one question that we must always ask is: "Are there one or more unwanted visitors in our collection of data?" For, if there are, we may never find a pattern that can account for invited and uninvited visitors alike. Watson and Crick, in fact, proudly proclaimed that, during the work that led to the DNA double helix, they ignored any data that did not fit their model. One purpose in generating a hypothesis is to tell us which data are relevant and which are not.

The flip side of the issue is that we can never be sure that we have all the data we need when generating a new hypothesis. I call these "missing guest" problems. It takes a particularly clever scientist to set the table for over a hundred guests when only thirty have RSVP'd. This is what Mendeleev did and why subsequent scientists consider him so smart: he set places in his periodic table for the handful of elements then known and left gaps for nearly a hundred elements that no one else suspected. Some of these elements would not show up to his party for over a hundred years! That's the kind of foresight that comes only with a clear understanding of how incomplete our knowledge is.

My other addition to the Wittes' ignorance curriculum has been to realize that each type of ignorance requires different methods to generate appropriate types of questions.[12] Simply asking the standard six questions—who, what, why, when, where, and how—does not suffice. For example, discovering hidden ignorance would seem to be particularly difficult. How can one uncover things we don't know we don't know?

Actually, scientists have been discovering unknown unknowns for as long as they have been performing controlled experiments. To set up a useful experiment, one must first have a hypothesis that clearly states what factors will affect an experimental system, in what way, and to what degree. One also defines what factors should have no effect. A "positive control" in a science experiment consists of testing factors known to affect the system in the same way as a new factor to be tested. A "negative control" is a factor that should not have any effect on the system. Science textbooks usually describe the role of controls as means to ensure the legitimacy of the experiment. A new chemical that one has hypothesized will lower blood pressure should have the same effect on blood pressure as drugs already known to lower it. Conversely, one also has to demonstrate that chemicals known to have no effect on blood pressure in fact have no effect in your own system on the particular day you are testing your new drug.

Controls are also one of the best means of uncovering hidden ignorance. A colleague of mine named Winston Brill told me the story of how he discovered one of the most potent and safest insecticides on the market today. He had a theory about what kind of molecules should behave as insecticides and synthesized dozens of variations of these. When he set up his experiment, however, he found that he had a few spare test tubes that he could fit into the automated test apparatus, so he added a few random chemicals out of bottles sitting around the lab. These random chemicals should have been his negative controls.

The results of the experiment were not what Brill expected. None of the chemicals he had designed and synthesized had any insecticidal activity (his hypothesis was wrong!), and one of the random chemicals that should have been a negative control worked wonders. Surprise! Brill now knew what he had not known before running his experiment, which was that there were unknown factors at work in controlling insect physiology. Following up these factors led to the invention of his new class of insecticides.

Serendipity is another fecund source of unknown unknowns, as another of my colleagues, Barnett Rosenberg, found out. All scientists start out their research with a particular goal in mind. The most successful often discover that the unanticipated phenomena that nature throws up along the way may be far more interesting and useful. That was Rosenberg's experience.

Many years ago, Rosenberg told me, he was studying the effects of pH on bacterial growth. One day, he noticed that his single-celled bacteria were beginning to divide without separating from each other. They started to form chains and clumps. No one had ever seen this before. He was fascinated. But he quickly determined that it wasn't due to alterations in the pH. Something he was doing that he didn't know he was doing was causing the effect. Months of work finally revealed that the acid in his media was eating away at the platinum electrodes he was using to measure the pH of his cultures. The dissolved platinum formed a new compound in the medium called cis-platin, and it was this that interfered with the cell division of the bacteria. Another colleague then suggested that anything that interfered with cell division might have potential as an anticancer agent, and, within a decade, cis-platin was to become one of the best-selling anticancer agents in the world. Serendipity had shown Rosenberg that there were things that he did not know he did not know that affected cell division, and, by discovering the nature of his unknown unknown, he was put on a path that would save millions of people's lives.

One key point about ignorance is apparent in both Brill's and Rosenberg's cases: you have to know what you do expect before you can be surprised by what you didn't expect. The unknown reveals itself, as Louis Pasteur once said, only to the prepared mind! You have to be prepared to recognize your ignorance if you are to benefit from it.

Now, clearly, the same methods that provoke hidden ignorance to become overt will not work to unveil things we think we know but we don't or things that we think we do not know but we do. Misknowns and hidden knowledge are revealed by different approaches. One is to question widely accepted but untested assumptions underlying current practice or theory. Would other assumptions (such as HIV works with cofactors) do as well or better? Another is to pay attention to anomalies, data that contradict the predictions of the dominant paradigm: Somehow information gets out of proteins and into the immune system in a usable form, so what's wrong with the way the Central Dogma is set up?

My favorite way to reveal misknowns is to turn things on their head. The Nobel laureate James Black has said that this technique is habitual for him. If someone tells him that the speed of light is constant, he'll immediately ask: "What would happen if it wasn't?"[13] The worst that happens in such instances is that one learns in the deepest way possible why physicists accept a constant speed of light; the best that can happen is that you discover that an entirely new approach to the field is possible. This is what Black discovered when he began investigating the causes of angina. According to the common medical wisdom of the time, angina resulted when the heart failed to get enough oxygen owing to blocked coronary arteries. Thus, Black's contemporaries busily set about finding ways to increase the amount of oxygen reaching the heart and, paradoxically, found that each intervention they tried made angina worse. Characteristically, Black turned the field on its head. What, he asked, would happen if we tried to *decrease* the amount of oxygen the heart needs to begin with? In order to test his idea, he had to invent a new class of drugs now known as "beta blockers." Not only did these drugs decrease the amount of oxygen needed by the heart; they also turned out to be very effective treatments for angina and other forms of heart disease. So questioning assumptions and asserting opposites can be very effective ways to reveal misknowns.

To discover things that we don't think we know but we really do (hidden knowledge) requires yet other methods. One of the most successful is to look to the history of your discipline. Many important insights, experiments, and ideas are discarded as science advances because they do not fit newly emerging fads—this is how I found out about the immunosuppressive effects of semen. The past may, therefore, illuminate the future. A similar ploy is to explore sources of information that are ignored by other people. Every discipline has its hierarchy of information sources. In science, we listen most carefully to those people at the top of the hierarchy—the people who have professorships at CalTech, MIT, Harvard, Yale, Stanford, and so on and the experts who publish articles in the most prestigious journals, such as *Science, Nature,* the *Proceedings of the National Academy of Sciences,* the *Journal of the American Medical Association,* and the *New England Journal of Medicine.* But what is a person to do when she has consulted all the bigwigs, read all the major papers and textbooks, and the problem remains unsolvable?

This is precisely what happened to physicians dealing with trauma patients around the globe during the 1980s. Unbeknownst to the public, three or four of every hundred patients admitted to a trauma center with major burns or traumatic injuries would develop an incurable infection. Antiseptics wouldn't work; antibiotics wouldn't work; even surgical removal of the infected tissue was often useless. Patients would suffer for months or even years with open, suppurating, stinking sores. Some would require amputation. Others would die. Two physicians, one in a clinic in the southern United States and another in Argentina, tried everything that modern medicine offered, failed, threw up their hands in despair, and began asking anyone and everyone for suggestions.

That is when they came on the unknown answer, which had been available for more than two thousand years.[14] In each clinic, a nurse, who had had the answer all along but who had not previously been consulted, suggested to the physician that the incurable wounds of the patients be slathered with honey. After all, that's what generations of mothers and grandmothers, shamans and witch doctors, had used to treat burns and wounds in the backwoods of the Deep South, the jungles of the Amazon, the rice paddies of China, and even the deserts of ancient Egypt. Anything used so widely for such a long period of time should surely have some therapeutic benefits.

As crazy as it sounds, the honey worked! Nine out of ten of the patients who were incurable by any other means would find their wounds fully healed within a few days or weeks. Large-scale clinical trials verified that honey, or a sugar-iodine paste, is far superior to any other treatment or set of treatments used for traumatic injuries, decreasing healing time, scarring, and hospital costs. This treatment is now also recommended for ulcerations associated with diabetes, sometimes in conjunction with maggots to eat away the dead and decaying flesh.

What stood in the way of honey therapy was that its practice existed among a group of disenfranchised people who were never asked about what they knew. Such self-imposed ignorance is based in social biases and prejudices that have no place in science or medicine. The ultimate test is not who knows what but whether what is known works.

Finally, the issue of taboo questions and unacceptable answers reveals that questioning is never a passive activity nor one that attracts the timid minded. To question an authority takes as much courage as to challenge a school bully. The risks are much the same and the mental bruises just as painful and long lasting as the physical ones. But not to question the experts incurs the same risks as refusing to fight for oneself: the powerful enslave the weak, the ignorant intimidate the knowledgeable. To learn without questioning is no different than brainwashing and makes the student an intellectual servant of those who place subservience above understanding.

Unfortunately, despite George Wald's admonition that it is questions, not answers, that make a scientist, nowhere in the entire science curriculum from preschool through postgraduate studies do students ever learn how to generate their own questions, let alone how to generate their own *insightful* questions! Where in all the curricula and the textbooks does one find any exercises that teach students how to ask or evaluate questions? Where in all the mess of facts is there any indication that there are questions that have never been answered and even questions that have never been asked? This is a travesty! We have put the cart before the horse by emphasizing answering over questioning. A person who can answer all the questions we already know the answers to, but cannot generate his own questions, can never become a scientist! I know! I went to college with such a person. He was a brilliant test taker and totally unimaginative. Such a person is in a position analogous to someone who can define and spell every word in the English language correctly but can't write a poem. Each has all the facts, but neither has developed the imagination to know what to do with them. Questioning builds imagination, for it takes us beyond known facts to face our ignorance squarely. What we do when we are faced with questions no one has previously asked and no one can presently answer is, ultimately, what makes or breaks a scientist.

Facing up to the importance of questioning should cause a revolution in education. The Wittes have been doing their part. I try to do mine. I incorporate problem finding and question posing into four aspects of every lecture and exercise. The first is to make certain that students understand what questions motivated the search for the knowledge that they learned that day. These questions provide a framework for them to hang their newly acquired knowledge on, a framework without which the facts pile up in a meaningless jumble. I also end every lecture by explicitly discussing the major questions that remain concerning the material we have just learned. Nothing excites my students more! They, like all bright individuals, want to make a mark on the world. Knowing that there are major questions that science has not been able to address entices their imaginations and provides them a model for how to form questions of their own. And I also tell my students to interrupt me at any time with questions that occur to them as I teach. "There are no stupid questions," I reiterate at regular intervals. I let them know that any question they have is surely shared by many others in the room too timid to ask, so by asking they will benefit themselves and others and, perhaps, prevent me from assuming something I think they know but they do not. And, fourth, I always tell them the story of Duncan Fisher's question about the Central Dogma and blood types so that they know that some of their questions might be so profound as to reveal whole realms of ignorance even the experts haven't glimpsed. Knowing how much there is yet to be known sets them free to imagine.

There is much more that surely needs to be done. Despite hundreds of broadly disseminated, formal ways to stimulate, teach, and evaluate problem-*solving* ability among students in every discipline imaginable, the literature on ways to stimulate, teach, and evaluate problem-*finding* and questioning ability is minuscule and has barely begun to be applied generally in any discipline.[15] Despite the fact that many psychologists are becoming more and more convinced that creativity in every field is more closely tied to problem finding than to problem solving, what we don't know about problem finding would fill many books. Like the drunk who lost his key in a dark alley but is found searching for it on his hands and knees under a lamppost at a well-lit corner because that's where he can see what he's doing, too many of us behave like experts, enslaved by the light and mortified by the dark. Odd, because the dark night of ignorance is where all the interesting unanswered questions are lurking!

So I regularly face away from the light, accept the resulting perceptual and conceptual blindness, and go ignorance hunting. It's always humbling to have to grope around on hands and knees like an infant again, and it's easy, too, for colleagues to make fun of your initially infantile progress, but it's also exhilarating when you learn your way around well enough to stand up and take your first teetering steps into the unknown. And it's the only way I've found to follow George Wald's advice to ask more and more meaningful questions. They're always the ones furthest from what we know.

Thus, I'm proud to say publicly and repeatedly: "I don't know!" I'm even prouder to do something about my ignorance, to dare the blank spots on the map of knowledge. Working just over the edge of the unknown is how I confirm that I'm really a scientist!

NOTES

1. Gerald Ames and Rose Wyler, *The Giant Golden Book of Biology* (New York: Golden, 1961), 2.

2. R. S. Root-Bernstein, "Mathematical Modelling of Biocheical Systems Using Petri Nets" (senior thesis, Department of Biochemistry, Princeton University, 1975).

3. R. S. Root-Bernstein, "Do We Have the Structure of DNA Right? Aesthetic Assumptions, Visual Conventions, and Unsolved Problems," *Art Journal* 55 (1996): 47-55.

4. James Watson, *The Double Helix* (New York: Atheneum, 1968); Francis Crick, *What Mad Pursuit* (New York: Basic, 1988); Robert Olby, *The Path to the Double Helix: The Discovery of DNA* (New York: Dover, 1994); Horace F. Judson, *The Eighth Day of Creation: Makers of the Revolution in Biology* (New York: Simon & Schuster, 1979); Matt Ridley, *Francis Crick: Discoverer of the Genetic Code* (New York: HarperCollins, 2006).

5. R. S. Root-Bernstein, *Discovering: Inventing and Solving Problems at the Frontiers of Knowledge* (Cambridge, MA: Harvard University Press, 1989), 29–31.

6. R. S. Root-Bernstein, "Amino Acid Pairing," *Journal of Theoretical Biology* 94 (1982): 885–94, "On the Origin of the Genetic Code," *Journal of Theoretical Biology* 94 (1982): 895–904, and "Protein Replication by Amino Acid Pairing," *Journal of Theoretical Biology* 100 (1983): 99–106.

7. R. S. Root-Bernstein and D. D. Holsworth, "Antisense Peptides: A Critical Review," *Journal of Theoretical Biology* 190 (1998): 107–19; I. Z. Siemion, M. Cebrat, and A. Kluczyk, "The Problem of Amino Acid Pairing and Antisense Peptides," *Current Protein and Peptide Science* 5 (2004): 507–27; J. Sibilia, "Novel Concepts and Treatments for Autoimmune Disease: Ten Focal Points," *Joint, Bone, Spine* 71 (2004): 511–17.

8. R. S. Root-Bernstein, *Rethinking AIDS: The Tragic Cost of Premature Consensus* (New York: Free Press, 1993).

9. L. Montagnier and A. Blanchard, "Mycoplasmas as Cofactors in Infection Due to Human Immunodeficiency Virus," *Clinical Infection and Disease* 17, suppl. 1 (1993): S309–S315; L. Montagnier, *Virus* (New York: Norton, 1999) and "A History of HIV Discovery," *Science* 298 (2002): 1727–28.

10. See the essay by Marlys Hearst Witte, Peter Crown, Michael Bernas, and Charles L. Witte in this volume.

11. R. S. Root-Bernstein, "The Problem of Problems," *Journal of Theoretical Biology* 99 (1982): 193-201; Root-Bernstein, *Discovering,* 56–66.

12. R. S. Root-Bernstein, "Nepistemology: Problem Generation and Evaluation," in *International Handbook on Innovation,* ed. L. V. Shavanina (Amsterdam: Elsevier, 2003), 170–79.

13. "Daydreaming Molecules," in *Passionate Minds: The Inner World of Scientists,* ed. L. Wolpert and A. Richards (Oxford: Oxford University Press, 1997), 124–29.

14. R. S. Root-Bernstein and M. M. Root-Bernstein, *Honey, Mud, Maggots, and Other Medical Marvels* (Boston: Houghton Mifflin, 1997), 31–43.

15. J. W. Getzels, "Problem-Finding and the Inventiveness of Solutions," *Journal of Creative Behavior* 9 (1975): 12–18, and "Problem Finding: A

Theoretical Note," *Cognitive Science: A Multidisciplinary Journal* 3 (1979): 167–72; M. A. Runco, "Problem Construction and Creativity: The Role of Ability, Cue Consistency, and Active Processing," *Creativity Research Journal* 10 (1997): 9–23; C. Christou, N. Mousoulides, M. Pittalis, D. Pita-Pantazi, and B. Sriraman, "An Empirical Taxonomy of Problem Posing Processes," *ZDM* 37 (2005): 1–27; T. Lewis, S. Petrin, and A. M. Hill, "Problem Posing—Adding a Creative Increment to Technological Problem Solving," *Journal of Industrial Teacher Education* 36 (1998): 1–30; R. F. Subotnik, "Factors from the Structure of Intellect Model Associated with Gifted Adolescents' Problem Finding in Science: Research with Westinghouse Science Talent Search Winners," *Journal of Creative Behavior* 22 (1988): 42–54; E. A. Silver, "On Mathematical Problem Posing," *For the Learning of Mathematics* 14 (1994): 19–28.

LESSONS LEARNED FROM IGNORANCE

The Curriculum on Medical (and Other) Ignorance

Marlys Hearst Witte, Peter Crown, Michael Bernas, and Charles L. Witte

The greatest single achievement of science in this most scientifically productive of centuries is the discovery that we are profoundly ignorant. We know very little about nature and we understand even less. I wish there were some formal courses in medical school on medical ignorance, textbooks as well, although they would have to be very heavy volumes.

—Lewis Thomas

What lessons can be learned from *ignorance* and particularly from *medical ignorance?* Lewis Thomas's novel idea for a course on medical ignorance struck a responsive chord. As an example, consider the state of ignorance about AIDS. After more than twenty-five years of fundamental discoveries, multiple clinical drug trials, and frustrating efforts at prevention, there has been little dent in the burgeoning global pandemic, and the prospect of effective vaccine development remains elusive. While much has been learned about AIDS, we still suffer from our ignorance. Perhaps an admission of ignorance, symbolized by a few blank pages in the section on AIDS in medical textbooks, might stimulate young minds to pursue new paths of investigation. The same approach holds for solid organ cancers, such as of the brain and the pancreas, where blank pages might more accurately reflect how little can be done practically to arrest tumor growth. Even artificial hearts and

organ transplants, the miracles of modern surgery and biomedical engineering, attest to the fundamental ignorance of heart disease and other
organ dysfunction. Indeed, surgery itself as a discipline is the ultimate
medical exercise in ignorance—removing organs and "mutilating" the
body. As the legendary surgeon John Hunter summed up several centuries ago: "It [operation] is like an armed savage who attempts to get
that by force which a civilized man would get by stratagem."[1] Although
an operation is often the best treatment currently available for many ail-
ments, it is at the same time a stark testimonial to a basic lack of understanding—namely, ignorance—of the underlying disease process and
an inability to prevent or arrest its progression by "natural" means. In
this setting, we reasoned that courses on medical ignorance would not
only enhance medical education but also be a potent stimulus for new
ideas and fundamental research. As the Curriculum on Medical (and
Other) Ignorance (CMI) evolved, the power and reach of the "ignorance
paradigm" has progressively revealed itself.[2]

THE PHILOSOPHICAL SHIFT FROM EPISTEMOLOGY TO NEPISTEMOLOGY

It is widely held in Western thought that knowledge and ignorance are
polar opposites, as if ignorance is a synonym of stupidity and, thus, an
insult.[3] This can lead to the impression that knowledge represents light
and progress and that ignorance is the enemy and an obstruction. We
have learned to fear ignorance if knowledge is to triumph and humans
are to advance. Quite the contrary. As Disraeli once wrote: "To be conscious that you are ignorant is a great step to knowledge."[4]

Philosophers debating epistemology for the past three millennia
have paired the study of knowledge with the study of ignorance, and, in
the real world, knowledge and ignorance are not irreconcilable polari-
ties. The Polish astronomer Copernicus's view was: "To know what we
do know and to know what we do not know, that is true knowledge."[5]

In fact, knowing and not knowing are intertwined and symbiotic.
Just as it takes knowledge to recognize ignorance, it takes an acceptance
of ignorance to face and inquire about what we do not know. As William
James noted: "Our science is a drop, our ignorance a sea."[6] From this
point of view, ignorance is not a void, but, rather, the fertile space of the
unknown that can provide the motivation to explore. Learning itself is
a continuing encounter with ignorance. In fact, the more we understand

something, the more we realize how little we know about it. As Pascal observed: "Knowledge is like a sphere, the greater its volume, the larger its contact with the unknown."[7]

Seen in this light, ignorance is a dynamic force in learning and research, and its topography shifts with inquiry. There are at least six lands within the domain of ignorance:[8] all the things we know we don't know (known unknowns); things we don't know we don't know (unknown unknowns); things we think we know but don't (errors); things we don't know we know (tacit knowns); taboos ("forbidden" knowledge);[9] and denials. Each of these lands is explored and charted in ignorance maps in the CMI (see fig. 1).

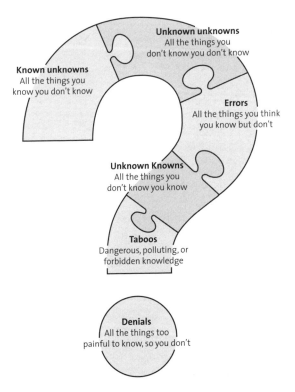

Figure 1. The ignorance map. (Based on M. H. Witte, A. Kerwin, and C. L. Witte, "Curriculum on Medical and Other Ignorance: Shifting Paradigms on Learning and Discovery," in *Memory Distortions and Their Prevention,* ed. M. J. Intons-Peterson and D. L. Best [Mahwah, NJ: Erlbaum, 1998], fig. 8.2, p. 137.)

The ignorance maps of a particular subject, for example, AIDS, will differ among the people creating them because of individual differences in their creators' previous learning and attitudes toward ignorance. However, the methods for working on the maps will be similar. Things we know we don't know, what Root-Bernstein refers to as "explicit ignorance,"[10] can be addressed by first questioning assumptions. If a problem has existed for an extended period of time, it means that experts probably don't have the right question. Root-Bernstein also suggests that we search for surprises by doing research that challenges expectations. Another approach is to focus on paradoxes because a well-defined problem will be embedded in a contradiction or a paradox. As Root-Bernstein points out: "The greatest scientists retain the child's wonder about the universe."[11] This suggests that the sublime can exist within what appears to be mundane.

THE INFORMATION SUPERHIGHWAY AND BYWAYS OF IGNORANCE

Society at large is in the midst of a profound paradigm shift as the Internet grows and proliferates. The information explosion has gradually transformed us into a knowledge culture, but a culture that stops short of wisdom. The Pew Internet and American Life Project found in its February–April 2006 survey that 73 percent of American adults, or about 147 million people, use the Internet and that 91 percent of these use search engines to find information.[12] *To Google* has become a verb meaning to search for information online using the Google.com search engine. In the United States alone, 91 million searches are performed on Google each day. When all the major search engines are included, the number increases to 213 million per day.[13]

While people turn more and more to the Web for information, and the amount of information available online increases dramatically, so too does the problem of evaluating the veracity of the information returned from an online search. For those who take comfort in the reliability of peer-reviewed papers, abstracts, and citations, the Internet gives instant access to many journals. Such access previously took time and resources to obtain if the journals one wanted were not in one's university library. (Google acknowledges this need with its "Google Scholar" in beta test in 2007.) But how does one evaluate other kinds of online search results, particularly when they come from a seemingly reliable source such as Wikipedia.org, the popular online encyclopedia? As of

July 2006, Wikipedia generated more traffic than MSNBC.com and the online versions of the *New York Times* and the *Wall Street Journal* combined, making it, as Stacy Schiff informs us, the seventeenth-most-visited site on the Internet. Despite its popularity, part of the problem with Wikipedia is, as Schiff points out, one of provenance, each article being created by an individual and then further edited by others who ostensibly know something about the topic: "The bulk of Wikipedia's content originates not in the stacks but on the Web, which offers up everything from breaking news, spin, and gossip to proof that the moon landing never took place. Glaring errors jostle quiet omissions."[14] That information of this caliber is so readily available on the Internet underscores the need to question both the information and its source. In his insightful *Scientific Literacy and the Myth of the Scientific Method,* the chemist Henry Bauer maintains that all knowledge, from the most subjective and unreliable to the most reliable and objective, passes progressively through a "knowledge filter" (see fig. 2). There are "all different degrees of probabilities and certainties that are changing always with time as we go from word of mouth to what's in the textbooks."[15]

The abundance of information now at our fingertips raises the issue of navigating and utilizing it. The management expert Peter Drucker advises: "Be data literate. Know what to know. . . . The challenges increasingly will be not technical but to convert data into usable information that is actually being used."[16] And the media philosopher Neal Postman expresses fear of a society of "information junkies" overloaded with "information which they don't know what to do with, [and] have no sense of what is relevant and what is irrelevant."[17] Further, Norman Myers suggests that the major environmental problems in the world are less a consequence of ignorance than of "ignor*e*ance," meaning that we choose to ignore problems. We choose not to study them, and, when we have the information, we choose to turn our backs on it.[18]

THE BENEFITS OF CHAOS AND FAILURE IN THE WORLD OF IGNORANCE

In a 1989 graduation address to the St. Louis University School of Medicine, we pointed out links among the unruly, boundless, negative spaces of ignorance, failure, and chaos as bright prospects for the future.[19] In *Chaos,* James Gleick describes the discovery of the revolutionary chaos

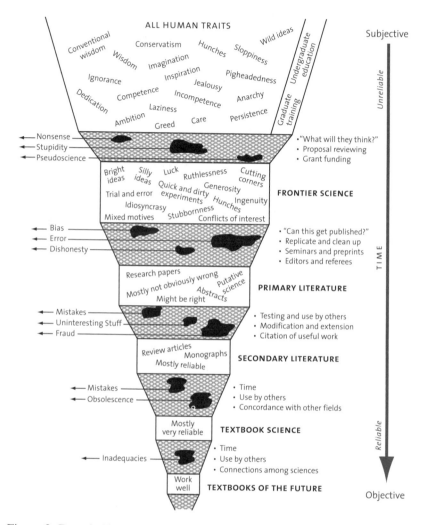

Figure 2. Bauer's "knowledge filter." (Based on H. Bauer, *Scientific Literacy and the Myth of the Scientific Method* [Urbana: University of Illinois Press, 1992], fig. 2, p. 45. Courtesy of Henry H. Bauer.)

theory by a physicist who devoted countless hours to watching meandering streams and gathering clouds.[20] Chaos theory hypothesizes that chaotic phenomena such as the atmosphere and turbulent fluids disobey the simple linear laws once thought to govern them. Whether a butterfly flapping its wings in Rio de Janeiro and creating a tornado in Texas or turbulent blood flow in an arterial aneurysm, recent research in nonlinear dynamics suggests that chaos itself may be desirable. Health, youth, and nondisease represent an ongoing chaotic disequilibrium poised for rapid response and adaptation. Aging is a condition of increasing stability where arrhythmias and other life-threatening disorders are more likely because the capacity to adjust is muted. Ironically, "pro-chaos" manipulations may, thereby, be useful treatment for a variety of diseases.

The relation of chaos and failure to success is seen in nonmedical fields as well. For successful corporations, the management guru Tom Peters advises "thriving on chaos" because those companies that recognize chaos also appreciate uncertainty and take the risks required to succeed.[21] R. S. Wurman, in his provocative *Information Anxiety,* examines failures:

> The winds of Puget Sound twisted, contorted, and destroyed the Tacoma Narrows Bridge, but also prompted urgent and exacting aerodynamic research that ultimately benefited all forms of steel construction. Beauvais Cathedral was built to the limit of technology in its day, and it collapsed, but subsequent cathedrals made use of its failure.
>
> Who's to know where any technology ends if its limits are not stretched? The machines of the world's greatest inventor, Leonardo da Vinci, were never built, and many wouldn't have worked anyway, but he was trying solutions where no man knew there were even problems. Clarence Darrow became a legend in the courtroom as he lost case after case, but he forced reevaluations of contemporary views of religion, labor relations, and social dilemmas.
>
> Edwin Land's attempts at instant movies (Polarvision) absolutely failed. He described his attempts as trying to use an impossible chemistry and a nonexistent technology to make an unmanufacturable product for which there was no discernable demand. This created the optimum working conditions, he felt.[22]

Largely since the insider-trading scandals and implosions on Wall Street, business and engineering schools have introduced courses like the University of Houston's "Failure 101" to analyze and rethink what happens when a great idea goes down in flames or a brilliant strategy fails.[23] Game players, whether chess masters or football stars, know that it's great to win but that learning is greatest after losing and then analyzing the reasons for losing. Physicians traditionally have "mortality and morbidity" conferences (the equivalent of "failure rounds") to pinpoint what went wrong in caring for patients who sustain unexpected complications or death and to discuss how to avoid undesirable outcomes. Whereas greater knowledge of the known can sometimes improve outcomes, more often such crucial knowledge remains to be discovered. Increasingly, critics of medical practice are examining the dynamics of mistakes in clinical decisionmaking so that performance and innovation can be enhanced.[24]

FROM KNOWLEDGE AND ANSWERS TO QUESTIONS AND LEARNING: A DIFFERENT APPROACH TO TEACHING

Lloyd Smith, the associate dean for medical admissions at the University of California, San Francisco, was once quoted as saying: "Our students should be freed from the stupefying excesses of fact engorgement, which threatens to convert them into floppy disks encoded for our present ignorance."[25] As medical educators, we must prepare physicians for the future. Yet medical students spend much of the basic-science years memorizing the knowledge of the day, much of which will be outdated shortly thereafter. (As Mark Ravitch has noted, putting things in perspective: "Textbooks of several generations ago were as large as the textbooks of today; they just contained different misinformation.")[26] Therefore, it is critical that medical students learn *how to learn, conscientiously and continuously,* over the course of a lifetime. In order to function in the uncertain terrain of medical diagnosis and treatment, students must be able to question skillfully and effectively; they must identify known unknowns, seek unknown unknowns, and question pat answers. Such focused expertise in learning, specifically in the identification and utilization of medical ignorance, is not a standard feature of current medical curricula. In order to help students become lifelong learners, we inaugurated in 1984 the CMI in the Department of Surgery as part of the Medical Student Research Program at the University of Arizona.[27]

The core content of this curriculum is the terrain of ignorance. How does one "teach" ignorance, that is, acquire the skills and attitudes to recognize and deal with it? Small children are prolific questioners. As they progress in school, however, most get the message not to ask questions, and universities and medical colleges too often reinforce this. In contrast, CMI demands and reinforces the questioning process. To help students relearn it (the "uncorking" of questions), we first solicit as many as possible from the student regarding a series of specific medical topics, for example, AIDS, breast cancer, gene therapy, obesity, organ transplants, stem cells, or artificial hearts. This releases inhibitions, leading to a barrage of questions. At this point, a dialogue begins using the classification of three generic types of questions. Type 1 questions are fundamental questions about basic biology, Type 2 questions are clinical or practical management questions about a specific patient or a specific medical problem, and Type 3 questions are those that go beyond the patient and relate to societal, economic, legal, and ethical issues. Then the student practices formulating questions of each type to encourage systematic questions of a wide-reaching variety. Many pedagogical references exist on questioning and inquiry, but most focus on the questions that the *teacher* should ask. In contrast, the CMI expects the *student* to ask the questions and to pursue them.

In summary, the basic content of the CMI is questions, and its method is questioning. Even though most physicians become accustomed to the role of questioner, many fear that an admission of ignorance may cause their patients or peers to lose respect for them. CMI endeavors to teach, discuss, and gain respect for the acknowledgment of ignorance.

SYLLABUS FOR THE CMI

A prime motivation for the CMI is to counterbalance the illusion of universities as "knowledge factories" with the reality that they are "ignorance commons," that is, ideal places for learning or discovering the unknown. This "reality" embodies the curricular reform suggested by Thomas as well as Pascal's enlarging sphere of knowledge. The more we know, the more we need to recognize what we don't know. To be overconfident with limited knowledge about poorly understood diseases is to be unwise, and, for a physician, this can even be dangerous. How, then, does one amalgamate these themes and ideas into a bona fide curriculum? First, we prepared a curriculum syllabus.[28] The CMI's goals in a nutshell are to gain an understanding of the shifting domains of ignorance and to learn the skills and attitudes necessary to navigate

them. Students learn to improve skills such as questioning and collab-
orating to recognize and deal productively with ignorance. Along the
way, positive attitudes and the values of curiosity, skepticism, humility,
optimism, and self-confidence are reinforced.

The Summer Institute on Medical Ignorance (SIMI) involves 20–25
percent of our medical students. Since 1987, it also incorporates approx-
imately 432 disadvantaged high school students (largely from underrep-
resented minority groups) and 135 kindergarten through twelfth-grade
science teachers. Each high school student is paired with a physician, a
basic scientist, or a medical student mentor. The program, symbolized
by a question mark, has students begin with questions at the start of
their projects and end with a new set of questions that remain unanswered at summer's close. The basic and clinical research experiences
are a full-time, hands-on, "brain-on" immersion in ignorance. Biweekly
seminars involve faculty members from the University of Arizona and
visiting professors, who present their own questions, probing what they
don't know and how questioning has stimulated research productivity
and clinical efficacy.

Thus, the first accredited medical school course on ignorance
encompasses a broad range of medical topics, such as AIDS, schizo-
phrenia, breast cancer, pain, liver cirrhosis, Alzheimer's disease, and
diabetes. Each student selects and "specializes" in a single subject, first
uncovering, through questions, what is known, and then delving into
what is not known. Ultimately the students assemble the material into
an oral and written final "ignorance report." They may also be assigned
a patient with a particular disorder, examine the decisionmaking processes in diagnosis and treatment, and identify the logical progression
of the questions already raised and other questions that should have
been. The central focus on unanswered questions requires effort and
thought. Thus, for example, to compose a sophisticated report on what
isn't known about breast cancer first requires formulating and organiz-
ing questions about what is currently known and accepted and then asking "what isn't known about breast cancer" (e.g., in terms of intricate
workings of the cancer cell, earlier and better diagnosis, genetic implications, and less destructive treatment).

In other exercises, visiting professors of medical ignorance offer
two seminars. The first is a traditional "knowledge" seminar in an au-
ditorium setting. The second may take place in La Residencia del In-
cógnito (the house of ignorance), the home of the nonprofit Ignorance

Foundation. Here, the visiting professors discuss what they don't know about the same subjects. The latter discourse is characteristically much livelier, with greater group participation and interchange. The contrast between lecturing on what is known and exposing what is not known, and the reaction of the students to enlightened ignorance, generates a very different kind of interactive experience.

TOOLS FOR TEACHING IGNORANCE

The guiding symbol and reminder is the ignorance map (see fig. 1 above), a large question mark with indistinct borders and abundant empty space. It is provided to the students to chart the lands of ignorance pertaining to a given topic, research project, or patient. Mapping gives a physical reality to the unknown and documents student progress in exploring its content and dimensions and "mining" its resources.

The ignorance log series is another key tool. Students submit weekly logs of personal daily questions along with their approach to finding the answers. Students can now fill out their daily ignorance logs on a secure Web site that saves their questions to a database so that the progression of their questioning can be reviewed and evaluated over time. Typically, they ask seemingly endless questions online, often pertaining to their personal life and professional future. (It is noteworthy that the Nobel laureate physicist Richard Feynman kept a daily diary of scientific questions that he called "things I don't know" and that resembles these ignorance logs.)

Since advances in medicine increasingly appear first in the newspapers or on the Internet, the CMI makes use of familiar media. A series of fictional but not improbable medical-scenario press releases have been developed to explore questions about other medical topics. In a creative-thinking exercise called "Grow Your Own Organ," the students analyze a sensational press release about a fictitious scientific breakthrough.[29] The press release begins: "3-month old baby girl dying from a rare metabolic liver disease received the gift of a new liver from her mother. Sources revealed that two months ago, a golf ball-sized portion of her mother's liver was removed and subsequently 'grown up' in tissue culture to a normal infant size in preparation for the historic operation." CMI students explain what the press release means to them, what their positive or negative reactions are, and what basic biological, clinical, and societal/moral/legal questions are raised (Types 1–3 questions). These questions are addressed, and the issues covered include how some questions can be approached and answered, how others are ongoing dilemmas, and how still others can be used as themes for teaching embryology, transplantation biology, or even medical ethics.

During "pondering rounds," students, faculty, and staff assemble to discuss the wide variety of things each has been contemplating in daily life, each telling their "ponders of the month" in less than a minute. Examples include: "Why do medical personnel sterilely scrub the site of lethal injection of condemned prisoners?" "Is there a rational basis for alternative medicine?" "How sexually active are octogenarians in nursing homes?" and, simply, "Where will my life take me?" Failure rounds focus on the errors committed and the disappointments experienced by the students and their mentors and how to correct them.

In "ignorance ward rounds," a complex patient is presented, and then questions are elicited from the audience, including the most knowledgeable senior faculty. The exercise is designed to bring the unanswered questions into the open for greater scrutiny. It was surprisingly difficult at first for medical students and residents because of their long-standing reluctance to admit that they have unanswered questions. Closer scrutiny revealed that they were actually more afraid to be tested on the questions raised. Once the students were assured that answers were not the point of the exercise, their questions poured forth. This illustrates how inhibiting the hospital teaching environment has become for questions yet how critical the process is for learning, unlearning, and correcting errors. Comparison of faculty-generated questions to those of residents and medical students reveals that many so-called stupid or naive questions that students and residents have suppressed are also uppermost on the minds of experienced professors and clinicians. This realization is a starting point for breaking down barriers between teacher and student so that medicine's limitations and unknowns can be frankly and productively examined.

Critical/creative-thinking workshops for teachers have encompassed diverse fields that range from music, history, and political science to business, civil engineering, and biomedicine. An undergraduate honors seminar at the University of Arizona covered what material science engineering is all about.[30] As medical faculty, we are less informed than most engineering undergraduates, yet we are able to raise student awareness of ignorance in their field and facilitate their questioning by "flaunting" our own naïveté. Our experiences have been similar regarding "AIDS and the Law" in the College of Law and "Anticipating the Future" in the College of Agriculture, the latter exploring possible future scenarios in the global environment and how ignorance and "ignoreance" are exerting unseen influences. Others have also experimented with ignorance-based instruction in higher education, for example, in journalism and psychology.[31]

Teachers practice converting one of their existing science lessons to a student question–based lesson. Examples of converted lessons during SIMI 2005 and 2006 covered topics such as surface tension, memory, eye anatomy, the metric system, continental drift, stem cells, and genetic engineering. In each transformation, the lesson begins and ends with unanswered questions from students.

A variety of conferences, symposia, workshops, and seminars have also been organized locally, nationally, and internationally around the theme of medical and other ignorance. As for AIDS, even the "establishment" now freely admits and articulates many uncertainties about the pathogenesis and treatment of this condition. To fill this gap, nearly two decades ago we organized two symposia (the first in Vienna in 1987, the second in Tokyo in 1989), under the auspices of the International Society of Lymphology, focusing on AIDS, Kaposi sarcoma, the lymphatic system, and medical ignorance.[32] In 1991 we featured AIDS as a topic addressed from the ignorance perspective by international experts in the field during the first International Conference on Medical Ignorance. Both the first and the second International Conference on Medical Ignorance (1991 and 2003) generated worldwide participation and interest.[33] The specific topics featured at the first conference, namely, AIDS, breast cancer, organ transplantation, and medical practice parameters (algorithms), remain "hot" areas of ignorance, along with other new ones (artificial hearts, gene therapies, etc.) introduced at the second conference.

It was not until 1993, however, that *Science* magazine, where many landmark articles are published, featured a special symposium entitled "AIDS: The Unanswered Questions." In this symposium, AIDS experts explored the many unresolved "known unknowns," including some that previously were erroneously claimed to be well understood.[34] More recently, in 2005, *Science* celebrated its 125th anniversary with a special issue on the 125 big unanswered questions in contemporary science, labeling its table of contents "Scientific Ignorance,"[35] and in 2006, the twenty-fifth anniversary of the discovery of AIDS, the journal focused on the unanswered questions surrounding this pandemic.[36] The *Journal of the American Medical Association* devotes an annual special issue—each one featuring a blank cover, symbolizing the unknown and unfulfilled—to AIDS.

Thus, ignorance as a pedagogical concept is easy to grasp and inexpensive to disseminate as an educational package. One can begin and end with students' questions, rather than (or in addition to) simply lecturing on what allegedly is known. Q^3 (Questions, Questioning, and Questioners) workshops and courses have been initiated under successive federal grants to assist Arizona's kindergarten through twelfth-grade science teachers in infusing curiosity, encouraging student questioning, and introducing students to "authentic inquiry"—how scientists actually practice science, proceeding from question to question. It is a practical and expeditious approach to introducing ignorance into science and interdisciplinary lessons, an approach that can transform classrooms into sorely needed, student inquiry–based curricula.

INTERNET AND MULTIMEDIA TOOLS ON IGNORANCE

As students, teachers, and physicians turn to the Internet and computer-based technology for tools to improve education and communication, we continue to develop software applications that support the CMI.[37] As mentioned previously, as part of the current Q^3 program, students submit their daily ignorance logs on a password-protected Web site. All logs are stored in a database for later evaluation. Final presentations by students and teachers include an "island of ignorance," which is a combined graphic and text report on a given topic. It includes initial questions required to get onto the island and the unknowns that remain after the original questions have been explored. We've experimented with brainstorming software such as Inspiration[38] and have been pleasantly surprised by creative, nonelectronic islands constructed from art materials and, once, from a chocolate cake! (Of course, it was devoured by all in attendance at the end of the presentation.)

The Questionator, programmed by the University of Arizona Web developer Michael Branch, is an interactive software tool for classroom use that encourages students to ask questions through participation in a gamelike activity. All students remain anonymous in the game, eliminating the possibility of embarrassment or intimidation, increasing the pace of individual questioning, and still allowing for positive reinforcement by the teacher. Student comments from the Questionator beta test have been very positive. The program will in the future be made available as an open source online. The Medical Ignorance Exploratorium and Collaboratory is a virtual, online meeting space where a number of people can work on a question together using live video and audio, shared files, images, documents, and shared desktops. It's based on the Macromedia Breeze server, making it low cost and widely accessible. We are now developing a Virtual Clinical Research Center. This will involve a national network of clinical research centers connected through state-of-the-art telemedicine technologies. Its purpose is to interconnect human and physical resources in the pursuit of collaborative approaches to answering medical questions and to give students and the general public the opportunity to interact directly with top researchers from around the world.

DOSAGE OF IGNORANCE

How much ignorance can a student or a teacher tolerate? Some medical students, often less than dazzling performers in the traditional medical curriculum, are superb at using ignorance-based learning. Often, they are curious questioners but are criticized for asking or are considered troublemakers or dissidents. Some may not excel at multiple-choice tests where pat answers are expected yet thrive on the uncertainty and ambiguity of an ignorance-based curriculum. For such students, a self-motivated discovery pathway could work for 95 percent of what they need to learn in medical school. Other students experience difficulty and discomfort with generating original questions where no simple answer exists. Yet this educational task is probably more akin to the demands of the thoughtful practice of medicine. Some students put it all together—balancing knowledge and ignorance and functioning comfortably at the interface. It may be wise, therefore, to offer a choice of learning styles to the students, but at least 5 percent of ignorance-based learning should be required to inculcate humility and resourcefulness, yet not ruin a high grade-point average!

EVALUATING IQ^3 AND Q VALUE

How should the ignorance curriculum be evaluated? How can acquisition of the desired attitudes, skills, and competencies to explore and value ignorance (the elusive ignorance quotient [IQ^3]) be determined? And how can short-term and future impact be measured? Clearly not by short-answer tests, but rather by the progression of questions (their number, depth, and fruitfulness), questioning, and questioners. The questions a student raises at first are compared with those at serial time points later in the course of exploration of a topic, and the progression is analyzed without directly requiring intermediate answers. This evolution of questions, together with what the student does about it, is the crucial test of a student's capacity to recognize and deal with ignorance and thereby acquire Q value, that is, grow the potential. Three education graduate students, Amanda Peterson, Heather Hopkins, and Deborah Schneider, initiated the development of a rubric to score the progression of understanding and sophistication demonstrated in each student's questions over time. We are also developing one that is meant to handle standardized, case-based assessment of questioning skills.

Pre- and post-self-assessment of attitudes, knowledge, and skills in recognizing and dealing with ignorance is also carried out. Highly significant changes have been documented.[39] Another source of data is student and participant feedback. Whereas some comments and suggestions reflect expected outcomes, others provide unexpected input, often in unsolicited letters. Here are some examples:

Medical Students

- One CMI graduate, now an emergency room physician, offers a litany of questions that come to mind as he practices emergency medicine: "Is the patient stable or unstable? Do I have an IV? Are they on oxygen? What medicines are they on? Are they important, etc." And he continues: "Medical ignorance abounds in the emergency department. It is precisely this active questioning and re-questioning that allows me to do a good job. It also has made me keenly aware of my limitations. Let's face it; one year of training no matter how good barely gains you admission to your own ignorance. I thank you for igniting the awareness of medical ignorance."
- Another graduate reports: "I remember telling you, I never asked questions in medical school. You gave me the courage to do so and it has shaped my entire clinical learning experience."
- Yet another indicated: "I learned that if our projects seem easy and we surely understand them, we are not really probing deeply;

it is only when we become perplexed by all the new discoveries unfolding before us that we can safely admit to being deep into the realm of research."

- Another student, now a successful researcher and board-certified anesthesiologist on a medical school faculty, writes: "I particularly appreciated the exuberant atmosphere, the emphasis on mistakes and the breaking away from 'safe science,' the potential fruit of a well-made mistake or the analysis of one that was unavoidable. The emphasis on ideas by stressing that medicine is a prisoner of a vast array of bias and dogma, that we are ignorant of what causes problems for our patients. We then are able to focus at least one eye somewhere over the horizon. . . . It is also non-elitist; bread and butter clinicians can be part of this reform and I'm happy to have joined the forces of the enthusiastically ignorant."

Undergraduate Students

- A material science engineering major offered an unsolicited essay called "Ignorance Is a Virtue, I Think." He wrote about studying a rock sample, saying: "There is no end to what possibly could be known. Compared to the amount of knowledge that exists, the amount that even the most intelligent human knows is infinitesimal."

Disadvantaged High School Students

- Many disadvantaged high school student research apprentices do not write well; some speak English poorly. Yet, to many, the impact of engaging the questioning process is clear. One concludes: "I found it amazing how much is going on that I didn't know about and I'm glad I had the chance to uncover it although it has raised many questions—this will make me go out and find out the answers and it will also raise many more which is the way you learn about life." Others expressed similar views, as the following excerpts make clear.
- "I benefited from the realization that without ignorance, humans would never advance as a civilization. The joys and pains that result from discovery would be lost forever if humans were omniscient."
- "Before this Summer I always hesitated to ask questions for fear

that I would look dumb. Now I know I was way off base. I've learned that the only way we can learn is through asking questions."

- "So I got over all the pride I had . . . started asking the Dr. my questions to understand things better. Once he saw that I was lost, he started explaining things to us better and we would ask more questions about the explanation. It was great. At that time I started to like coming to work because I knew I wouldn't be some random kid following around a doctor and his staff, . . . that I was part of that staff. It was really fulfilling."

- "If someone is working on a very difficult research project and has dug themselves into a hole, an onlooker may present a question that is very valuable. . . . It may help the person doing the research by opening up a new area of thought. Perhaps something they have not thought of before will become apparent to them."

Kindergarten through Twelfth-Grade Science Teachers

- A second-grade teacher said that she now was able "to develop the skills in asking questions strategically and learning to motivate my students to do the same while eliminating the negative feelings that hold us back from asking such important questions."

- Another said: "SIMI really got me thinking about questions, the importance of student-based questions in directing classroom flow. I had never really considered student-centered questions versus teacher-centered."

- Yet another reported: "[After this summer], I am more skeptical . . . more encouraged to let the students do brainstorming. . . . I have seen timid students blossom. . . . They are now more assertive and more willing to take chances. Students have shown lots of leadership and they learn to figure out things for themselves."

Thus, evaluation extends beyond participant performance and ignorance products to examining the empowerment process and the chain reaction that is set into motion by the questioning curriculum.

IGNORANCE REFLECTIONS

True "experts" acknowledge more ignorance than those less versed. As they enter the terra incognita, they take risks and ask questions that multiply and fertilize. Richard Feynman summarizes it well: "When a scientist doesn't know the answer to a problem, he's ignorant. When he has a hunch as to what the result is, he's uncertain. And when he's pretty darn sure of what the result is going to be, he is in doubt still."[40]

Alfred P. Sloan Foundation President Ralph Gomory emphasizes the need for teaching about ignorance: "We are all taught what is known, but we rarely learn about what is not known and we almost never learn about the unknowable. That bias can lead to misconceptions about the world around us. . . . Science excels in creating the artificial and controllable increasingly knowable world . . . [yet] two limitations may constrain the march of predictability. First, as the artifacts of science and engineering grow ever larger and more complex, they may themselves become unpredictable. . . . And second, embedded within our increasingly artificial world will be large numbers of complex and thoroughly idiosyncratic humans."[41] As the Information Age shifts into the next phase, the real lesson of this past century is that we never did, never will, and never should have pat answers. The coming Age of Ignorance and the "ignorance explosion" should prove to be a new age of enlightenment, where ignorance, failure, chaos, and unanswered questions emerge from the darkness and fulfill their potential for enriching lives and expanding future possibilities.

NOTES

Supported by funds from the American Medical Association (AME-ERF), National Institutes of Health (HL07479—Short-Term Training in Professional Schools, R25RR10163—NCRR Minority Initiative: K–12 Teachers and High School Student Program, NCRR Science Education Partnership Awards R25RR 15670 and 22720), Eisenhower Math and Science Education Act, Arizona Board of Regents and Dean's Funds, University of Arizona College of Medicine. Opinions expressed herein do not represent the views of the sponsoring agencies.

1. J. Hunter quoted in J. Kobler, *The Reluctant Surgeon: A Biography of John Hunter* (Garden City, NY: Doubleday, 1960), 108.

2. M. H. Witte, A. Kerwin, and C. L. Witte, "Curriculum on Medical and Other Ignorance: Shifting Paradigms on Learning and Discovery," in *Memory Distortions and Their Prevention,* ed. M. J. Intons-Peterson and D. L. Best (Mahwah, NJ: Erlbaum, 1998), 127–59.

3. A. Kerwin, "None Too Solid, Medical Ignorance," *Knowledge: Creation, Diffusion, Utilizations* 15 (1993): 166–85.

4. B. Disraeli, *Sybil; or, The Two Nations* (Oxford: Oxford University Press, 1845).

5. Traditionally attributed to Copernicus.

6. W. James, *The Moral Equivalent of War and Other Essays,* ed. with an introduction by John K. Roth (New York: Harper & Row, 1971), 82–83.

7. B. Pascal quoted in M. H. Witte and C. L. Witte, "Epilogue: Beyond the Sphere of Knowledge in Lymphology," in *Cutaneous Lymphatic System,* ed. R. Ryan and P. S. Mortimer, special issue, *Clinics in Dermatology* 13 (1995): 511.

8. Kerwin, "None Too Solid."

9. R. Shattuck, *Forbidden Knowledge* (New York: St. Martin's, 1996).

10. R. Root-Bernstein, "Problem Recognition and Invention: An Idiosyncratic View of Nepistemology" (paper presented at Q^3 Workshop 3, Medical Ignorance Collaboratory, 20–21 July 2001).

11. Ibid.

12. See "Internet Activities," http://www.pewinternet.org/trends/Internet_Activities_7.19.06.htm.

13. See Danny Sullivan, "Searches per Day," 20 April 2006, http://searchenginewatch.com/showPage.html?page=2156461.

14. S. Schiff, "Know It All: Can Wikipedia Conquer Expertise?" *New Yorker,* 31 July 2006, 36, 43.

15. H. Bauer, *Scientific Literacy and the Myth of the Scientific Method* (Urbana: University of Illinois Press, 1992), 45.

16. P. F. Drucker, "Be Data Literate: Know What to Know," *Wall Street Journal,* 1 December 1992, Eastern ed., A16.

17. N. Postman, *MacNeil/Lehrer Newshour,* PBS, 25 July 1995.

18. N. Myers, "Environmental Unknowns," *Science* 269 (1995): 358–60.

19. M. H. Witte, "Medical Ignorance, Failure, and Chaos: Bright Prospects for the Future," *Pharos* 54 (1991): 10–13.

20. J. Gleick, *Chaos: Making a New Science* (New York: Viking Penguin, 1987).

21. T. Peters, *Thriving on Chaos: Handbook for a Management Revolution* (New York: Knopf, 1987).

22. R. S. Wurman, *Information Anxiety 2* (New York: Doubleday, 2001), 273.

23. R. Johnson, "To Pass This Course, the Students Have to Try Their Hardest to Fail," *Wall Street Journal,* 16 January 1989, B1; D. Blum, "Risk-Taking Encouraged: In 'Failure 101,' U. of Houston Engineering Professor Offers an Innovative and Creative Approach to Design," *Chronicle of Higher Education,* 11 April 1990, A15.

24. A. Wu, S. Folkman, S. McPhee, and B. Lo, "Do Houseofficers Learn from Their Mistakes?" *Journal of the American Medical Association* 265 (1991): 2089–94.

25. L. Smith quoted in *Chronicle of Higher Education,* 20 June 1990, B1.

26. M. M. Ravitch, *Second Thoughts of a Surgical Curmudgeon* (Chicago: Yearbook Medical, 1987), 126.

27. M. H. Witte, A. Kerwin, and C. L. Witte, "Seminars, Clinics, and Laboratories on Medical Ignorance," *Journal of Medical Education* 63, no. 10 (1988): 793–95; M. H. Witte, A. Kerwin, and C. L. Witte, "Curriculum on Medical Ignorance," *Medical Education* 23 (1989): 24–29; M. H. Witte, C. L. Witte, and D. L. Way, "Medical Ignorance, AIDS-Kaposi Sarcoma Complex, and the Lymphatic System," *Western Journal of Medicine* 153 (1990): 17–23 (featured in the accompanying editorial, R. H. Moser, "Ignorance: Inevitable but Invigorating," *Western Journal of Medicine* 153 [1990]: 77, and in *International Medical News* 23 [1990]: 3); C. L. Witte, A. Kerwin, and M. H. Witte, "On the Importance of Ignorance in Medical Practice and Education," *Interdisciplinary Science Reviews* 16 (1991): 295–98; C. L. Witte, M. H. Witte, and A. Kerwin, "Ignorance and the Process of Learning and Discovery in Medicine," *Controlled Clinical Trials* 15 (1994): 1–4.

28. M. H. Witte, A. Kerwin, C. L. Witte, J. B. Tyler, A. Witte, and W. Powel, *The Curriculum on Medical Ignorance: Coursebook and Resource Manuals for Instructors and Students* (1989; rev. ed., Tucson: University of Arizona College of Medicine, May 1996).

29. Witte, Witte, and Kerwin, "Ignorance and the Process of Learning and Discovery in Medicine."

30. M. H. Witte, leader, "Spirit of Inquiry: Some Perspectives on Ignorance" (undergraduate honors seminar, University of Arizona, 29 August 1996).

31. S. H. Stocking, "Ignorance-Based Instruction in Higher Education," *Journalism Educator,* Autumn 1992, 43–53.

32. M. H. Witte, ed., "AIDS, Kaposi's Sarcoma, and the Lymphatic System: The Known and the Unknown," special issue of *Lymphology* 21 (1988): 1–87; M. H. Witte and T. Shirai, eds., "AIDS, Other Immunodeficiency Disorders, and the Lymphatic System: Pathophysiology, Diagnosis, and Immunotherapy," special issue of *Lymphology* 23 (1990): 53–108.

33. First International Medical Ignorance Conference, University of Arizona, 14–16 November 1991; Second International Medical Ignorance Conference, University of Arizona, 19–20 July 2003.

34. "AIDS: The Unanswered Questions," special section of *Science* 260, no. 5112 (28 May 1993): 1253–1396.

35. D. Kennedy and C. Norman, "What Don't We Know?" *Science* 309, no. 5731 (1 July 2005): 75.

36. "HIV/AIDS: Latin America and Caribbean," special section of *Science* 313, no. 5786 (28 July 2006): 405–541.

37. Most of these are summarized on www.medicalignorance.org.

38. Available at http://inspiration.com.

39. Witte, Kerwin, and Witte, "Seminars, Clinics, and Laboratories on Medical Ignorance."

40. R. P. Feynman, *What Do You Care What Other People Think?* (New York: Norton, 1988), 245.

41. R. Gomory, "The Known, the Unknown, and the Unknowable," *Scientific American* 272 (June 1995): 120.

Economics and the Promotion of Ignorance-Squared

Herb Thompson

For the past fifty years, there have been two developing trends at the core of the textbook neoclassical economic theory that is passed on to students. The first is that economists have become increasingly engaged in the formation of compelling, mathematically elegant hypotheses with little interest in their policy implications. The second is the reluctance of mainstream economists and their students to engage in conversations with alternative paradigmatic schools of thought (e.g., feminists, Marxists, or proponents of alternative economic models such as institutionalists or post-Keynesians). Because of this, economics (the mainstream theory taught in both secondary schools and universities) has become anti-intellectual and the promoter of an insidious form of ignorance, which I label *ignorance-squared. Ignorance-squared,* in this context, is meant to suggest a purposeful lack of concern regarding the awareness of what it is that one doesn't know.

In January 1996, a group of "heterodox economists" made a presentation to the Publishing Committee of the American Economics Association responsible for the *American Economic Review,* the *Journal of Economic Literature,* and the *Journal of Economic Perspectives* (with little effect thereafter, I might add!). Anne Mayhew, the editor of the *Journal of Economic Issues,* speaking before the committee, cogently pointed out that a small group of economists had "captured" the most prestigious journals of the discipline and, subsequently, promoted mathematical complexity at the expense of social, political, and economic issues incorporating "history, institutions and power." Further, the prestige of the association and the journals was unashamedly used "to narrow the discipline, to reward the excessive technical training of the prestigious graduate schools, and to stifle the advance of heterodox approaches to economics" (Mayhew 1996, 1–2).

What follows is a discussion of some of the content, causes, and effects of Mayhew's articulated frustration with the discipline's promotion of, and its consequent limiting of, economic knowledge.

KNOWLEDGE, IGNORANCE, AND IGNORANCE-SQUARED

Methodologically, economists are primarily concerned about the question, "How can one tell whether a particular bit of economics is good science?" (Hausman 1989, 115). Herein, I pursue a different, sociological question, not how we come to know what it is that economists don't know, but why it is the case that mainstream economists *don't want to be aware of what it is that they don't know.*

One of the streams of awareness being tracked in this collection of essays is that "the greatest achievement of science . . . is the discovery that we are profoundly ignorant; we know very little about nature and understand even less" (Kerwin 1993, 174). However, the beneficial aspects of this type of ignorance are lost in the much more insidious cultivation of, and purposeful production of, ignorance in the classroom. This is done through narrow pedagogy, the confining of research parameters, and constraining the production and presentation of non-neoclassical economic knowledge.

Training in textbook economics and economic research systematically fosters ignorance-squared in that students and researchers are shielded from any acquaintance with problems outside the domain of successful puzzle solving. The situation that occurs in classrooms is characterized as "purposeful ignorance" by Paul Heltne (in this volume). According to Heltne, it is the time "when you are made to feel that it would be foolish to ask questions about derived conclusions or about basic assumptions." Economics students are regularly confronted with this state of affairs. The curriculum is always crowded with the positive heuristic of neoclassical economics, as if there is always too much to teach. There is never time for reflection, for perspective, for the cultivation of awareness, and, most important, for the presentation of other contentious viewpoints, much less for the knowledge produced outside the disciplinary boundaries. When neoclassical economists restrict their discourse as well as their students' ability to engage with others of the same or related specialties, then ignorance-squared, in the manner put forward by Ravetz (1993), is enhanced.

We are all ignorant in a variety of ways, to various degrees, with respect to specific issues, problems, and questions. In fact, it is the increasing awareness of our ignorance of what there remains to know that is most special about the learning process and that is the focal point of this volume. A taxonomy of ignorance by Smithson (Smithson 1989, 9; and Smithson 1993, 135), outlined below, remains most helpful in suggesting the variety of forms (the reader should also note the taxonomies in this collection of essays provided by Berry and by Witte, Crown, Bernas, and Witte, the latter's efforts being very similar to that provided here):

- All the things of which people are aware they do not know (the most recognized "humble" form of ignorance described by Paul Heltne);
- All the things people think they know but do not (ignorance based on error); this form of ignorance comes close to what Heltne calls "ignorance-masquerading-as-certain-knowledge";
- All the things of which people are not aware that they do, in fact, know (intuition);
- All the things people are not supposed to know but could find helpful (taboo);
- All the things too painful to know (psychological suppression of memory); and
- All the things of which people are not aware that they do not know (ignorance-squared).

Of particular interest for our purposes is the final category. There is nothing necessarily negative about the fact that we proceed through life unaware of most of what there is to know. In fact, the humility of recognizing that fact may, indeed, be the beginning of knowledge. What I argue, however, is that neoclassical economists promote ignorance-squared.

There are many obvious reasons why this promotion exists, some of which are intuitive. For instance, given the search costs combined with specialization, there is only so much time to devote to methodological issues. Therefore, the dominant paradigm will draw the attention of most scholars. Moreover, the more a system (of thought) is entrenched, and the longer the time it has been operating, the more difficult and expensive it becomes to change it (Collingridge 1980). Likewise, the more a person has invested in the training required to be admitted to the neoclassical coterie, the more it is in that person's interest to prevent the depreciation of knowledge threatened by alternative methodological perspectives. Another reason may be the specific intellectual appeal of elegantly constructed mathematical neoclassical axioms. Given the conceptual apparatus of ignorance-squared, we can now examine the production of economic knowledge that incorporates, simultaneously, the production and the promotion of ignorance.

NEOCLASSICAL ECONOMICS: MARKETS, EFFICIENCY, AND SELF-INTERESTED RATIONALITY

The common view in economics remains embedded in the anachronistic philosophy of positivism. That is: "Scientific theories are derived in some rigorous way from the facts of experience acquired by observation and experiment. . . . Personal opinion or preferences and speculative imaginings have no place in science. . . . Scientific knowledge is reliable knowledge because it is objectively proven knowledge" (Chalmers 1982, 1). The dominant form of economics in the discipline, known generally as "neoclassical" economics, is made up of the hard core of orthodoxy that both derives from and reinforces the self-perception of Western civilization as a domain of autonomous, self-interested, rational individuals. This ignorance-squared is advanced in a discipline of economics that is largely ethnocentric and viewed from an Anglo-Celtic perspective. Economics texts ignore the content of economic thought as it has been developed in Germany, France, Italy, or Japan and, even more so, largely ignore the Arab or Islamic world, China, and the Indian subcontinent, just to take a few examples. The assumption must be either that these peoples didn't think about economic issues or that their thoughts are irrelevant to economics as we know it. The problem for students is that this is pushed to the tautological extreme by suggesting that economics as we know it is all that we need to know. Further to the point of ignorance-squared is that there are many and various schools within the discipline that are marginalized, such as the Austrian, institutionalist, and post-Keynesian. All these marginalized schools take a clear antipositivist position, but economists as a whole remain largely impervious to their antipositivist critique (Caldwell 1982). By controlling the development of economic knowledge, a neoclassical economic definition of reality is provided and made self-evident for the perusal of civil society. Given this unalterable definition of reality, all forms of economic policy are then logically placed within a "market" context (Samuels 1991).

This "market" has become the dominant metaphor, structuring and constraining the purposive behavior of the rational, self-interested, all-knowing, optimizing individuals of which it is constituted. Economic activity outside the market or the realm of exchange remains outside the realm of "relevant knowledge" and, therefore, solidly within the domain of ignorance-squared. For instance, reviewing, assessing, and suggesting improvements to subsistence agriculture in Pacific Island nations has been mainly left to anthropologists because of the explicit exclusion of nonmarket economic activity from the required knowledge to be found within the discipline of neoclassical economics. From the market metaphor, it is a short step theoretically, and quantitatively, to establish the mechanisms and principles of "competition" and "efficiency" (Negishi 1962). Competition (in the ubiquitous form of supply and demand) is to the market as the law of gravity is to the environment (Koppl 1990). And, from this, the mechanical equilibrium and harmony promoted in economic models of the environment are both achieved through an unfettered operation of competitive markets, with little, if any, government direction (Arrow and Hurwicz 1958).

In the classroom, students are usually invited to view a number of two-dimensional, ahistorical diagrams showing the reality of the economy as a mechanical entity at rest, hit periodically by exogenous shocks, but inevitably returning to equilibrium through the operation of the competitive gadgetry found in every introductory textbook. This view of a mechanical order has its roots in the tales told by Adam Smith, who, in turn, drew his enlightenment from the likes of René Descartes and Isaac Newton. The student need only open the textbook to discover an elegant world of individual agents engaged in the constrained pursuit of their self-interest through choices made in numerous interrelated markets (money, land, labor, capital, and goods and services). The purposeful effect is to uncouple the economy from both its sociocultural and its political life-support systems and to confer on practitioners a self-determining reality of their own. In this sense, economists spend much of their time imagining perfect worlds, uncluttered by political, social, or cultural messiness. The disturbing fact is that ignorance-squared is promoted by placing real processes outside the field of thought structured by logical, conceptual, a priori structures.

Perhaps Smith's most famous tale is exemplified in this often-quoted passage: "Every individual . . . endeavors as much as he can . . . to employ his capital . . . in such a manner as its produce may be of the greatest value, he intends only his own gain, and he is in this, as in many other cases, led by an invisible hand to promote an end which was no part of his intention. Nor is it always the worse for the society that it was not part of it. By pursuing his own interest he frequently promotes that of the society more effectually than when he really intends to promote it" (Smith 1910, 400). The importance of what was meant to be a commonsense assertion (irrespective of whether the reader agrees with it) is that that assertion forms the foundation of the hard-core belief system in orthodox economics. Starting with this assertion, economists have spent much of their time, for the past two hundred years, imagining a rigorous and elegant world. This world of "perfect competition" is presented as the positive heuristic of neoclassical economics.

In more contemporary orthodox terminology, the essential point of Smith's vision is that, in certain circumstances, economic agents need give no heed to the wider consequences of their actions. In pursing their own selfish objectives, they will, in the end, serve society's best interests. Just as the generators of a hydroelectric plant harness the power of falling water to a desirable end, the invisible hand of the market will harness the power of our base self-interest to society's benefit.

A great deal of intellectual effort has been expended this century in expanding on this simple idea and identifying precisely in what sense, and how, markets can deliver a socially desirable outcome. It can be shown that such a "perfectly competitive" economy will not only deliver economic efficiency but also disclose the most socially desirable results.

The connection between efficiency and social desirability is a policy beacon of orthodox economics. That is, "overall economic efficiency" requires and implies that the conditions of "technical efficiency" (which exist when all products are being produced with the minimum possible inputs), "efficiency in production" (which presumes technical efficiency above and elicits the situation when no further swapping of inputs between the lines of production will see an unambiguous improvement in output), and "efficiency in exchange" (all trades that will improve the well-being of any consumer without lessening that of another) are considered to have been met. This caps the neoclassical conception of economic efficiency since all opportunities for the improvement of any particular agent's position while leaving others no worse off have been exhausted.

As already intimated, economic efficiency in this rather special sense can be delivered by the operation of self-interested, rational individuals in a perfectly competitive market economy. While there are many permutations and combinations of this model, subject to continual refinement and development, it is sufficient for our purposes to focus on the four central requirements of perfect competition.

First, it is essential that all relevant information is available to all agents. Most important, it must be assumed that no single firm has an advantage (say, in the form of a secret and superior technology), that consumers are fully informed about the quality of all products, and that all the prices at which both products and inputs can be purchased are known. In short, there is no misinformation and no lack of information. No agent is excluded from information that would cause his or her behavior to change.

Second, a basic behavioral assumption requires each agent to be both self-interested and means/ends rational in the pursuit of that self-interest. This implies that consumers will pay the lowest price for any given good, that firms will use the best available technology, and that all agents will seek to maximize net benefits. This assumption, along with the first, ensures technical efficiency, not to mention efficiency in production and exchange. The "rational" consumer of the mainstream economist is a working assumption that was meant to free economists from dependence on psychology (Simon 1976, 131; Tversky and Kahneman 1987). The dilemma is that the assumption of rationality is often confused with, and regularly presented as, real, purposive behavior. In fact, the living consumer and producer in historical time routinely make decisions in undefined contexts. They muddle through, they adapt, they copy, they try what worked in the past, they gamble, they take uncalculated risks, they engage in costly altruistic activities, and they regularly make unpredictable, even unexplainable, decisions (Sandven 1995). In other words, they operate in a framework of ignorance, all assumed away by the logical constructs of neoclassical economists.

Third, it is assumed that all agents are price takers and act in an independent and noncollusive manner. In other words, all firms and consumers take the prevailing market prices as given and either make no attempt to influence them or simply could not even if they were to make an attempt to do so. Price taking is often ensured by assuming that there are sufficiently large numbers of buyers and sellers in every market such that no individual agent has any influence over any market price and that the best those buyers and sellers can do, therefore, is simply accept the prevailing prices. This assumption is central to delivering the optimal conformity of production and exchange under perfect competition and, thus, in turn, overall economic efficiency.

Fourth, and most significant for comprehending the neoclassical approach, it must be assumed that the market prices of all goods and services fully reflect all the costs and benefits of their production and consumption. This assumption means that, in the model of perfect competition, there are no uncompensated effects. Uncompensated effects arise when there are affected third parties not involved in the market transaction between producer and consumer. As a result of the production and/or consumption of the product, these third parties suffer some adverse consequence or obtain some benefit for which neither have they been paid nor are they required to pay. In the orthodox perspective, these uncompensated effects make up *the* environmental problem. The relevant markets do not exist. Indeed, in many cases, they cannot exist, as, for instance, when there are significant intergenerational effects.

THE ENVIRONMENTAL PROBLEM: MARKETS, EFFICIENCY, AND SELF-INTERESTED RATIONALITY

Consider the rather well-worn textbook example of a river in which, by tradition, a group of people enjoy fishing, recreational, and water rights. Suppose that an industry of many firms establishes its operations upstream and that part of its production process requires the use of water from the river. Further suppose that, while this industry's use of the river leaves the quantity of water unaffected, it does mean a deterioration in its quality, which affects some or all of the fishing and recreational activities of people downstream. The impacts can be in the form of a reduced commercial fish catch, less pleasant swimming or boating conditions—in fact, anything that makes the original users of the river feel worse off in their own estimation.

In this instance, we have a party adversely affected by the production processes of an industry for which, if economic efficiency is to obtain, they should be compensated but for which they will not be if it is left to the market. A market for the sale of the waste-absorption services of the river will not form automatically, as, say, a market for land will, since the relevant property rights may not be formally defined and the affected group will in all likelihood be too large and diverse to form an organized and effective body to deal with the industry. There is, in effect, a missing market. So long as the impact on the downstream river users remains uncompensated, an economic impact is not being registered in the price system. Under otherwise competitive conditions, the price at which the industry sells its product will not properly reflect the costs of producing it. There will be what is called a negative externality. The private costs to the industry for the production of its product are less than the total social costs of its production processes, which take into account both the industry's private costs and all uncompensated costs. But, for the economist, the issue is whether they can be compensated, not that they actually are compensated. Again, real processes are allocated to the realm of ignorance-squared, and the logical model takes precedence in analysis.

In the presence of negative externalities, the conditions for economic efficiency are violated. The crucial equality between costs and prices will not apply. The price for the industry's output will be too low and the production of its output too high to fully reflect the costs of production to society as a whole. There will be a misallocation of resources toward the upstream industry and away from all other productive activities; and the downstream users will have been made worse off with no compensation.

Static Linearity versus Nonlinear Complexity

Unlike the reality of ecologists and biologists, which is largely based on nonlinear systems, that of economists is predominantly constructed on linear foundations. As one might expect, problems must arise when orthodox economists bring these finely honed tools, normally used in generating models of perfect competition, to bear on understanding and explaining ecological complexity (Drepper and Mansson 1993, 44).

Texts that are primarily concerned with environmental economics reveal a crippling defect: they are, in most cases, envisaging a virtually static natural world fluctuating around a stable equilibrium (Drepper and Mansson 1993, 45; Kirman 1989). This is so even though, as Hutchison has insisted, "the assumption of a tendency towards equilibrium implies . . . the assumption of a tendency towards perfect expectations, competitive conditions and the disappearance of money" (Hutchison 1938, 107).

Although the quest for precise predictions of the effects of different policies is an ingrained feature of much economics, economists quickly discover the limits inherent in nonlinear systems. This is because the time horizon for reasonably reliable predictions grows much more slowly (logarithmically) than the corresponding demand for increased precision in the specification of the current state (Drepper and Mansson 1993, 50; Norgaard 1989). Several weeks before the 1929 stock market crash, the respected Yale monetary theorist Irving Fisher pointed out confidently that "stock prices have reached what looks like a permanently high plateau" (Coleman 1998, 64). Each new generation of students in economics gains amusement from the dated forecasts of its forebears and then proceeds to purposefully create its own realm of ignorance-squared.

In a nonlinear framework such as an ecological system, change is not proportional, and the components cannot be reduced or analyzed in terms of simple subunits acting together. The resulting properties can often be unexpected, complicated, mathematically intractable, or chaotic and in continual disequilibrium (May 1976).

There is, therefore, within the model of perfect competition a tension between the fundamental behavioral assumption of rational self-interest on the part of economic agents and the conditional requirement that there be no uncompensated effects. Just as the competitive process can produce an efficient outcome when the necessary conditions prevail, the competitive process can lead to the violation of one important set of those conditions. Firms seeking a competitive advantage over their rivals may well find that the way to reduce the price at which they can sell their product is to increase the proportion of their costs borne as externalities by uncompensated third parties.

The Economic Determination of Value

The first step for the economist, when analyzing issues or problems with respect to the environment, is to isolate all the measurable costs and revenues. The use of language to assist in this transformation of nature into marketable commodities helps to ideologically construct a new reality. "Exchange values" and "capital stock" require little in the way of discursive maneuvering to lead an audience to a discussion of new jobs, increased exports, and economic growth. These are powerful concepts, and they are regularly used contemptuously against those talking about the preservation of a threatened species or walking trails for urban dwellers.

With the advance of environmental economics as a subdiscipline, the neoclassical economist proceeds from a simplistic emphasis on direct market values to a more subtle incorporation of "nonmarket" values. In this instance, economists make the argument that they are prepared to assist their verdure opposition by establishing relevant methods of both analysis and colloquy. This shift is analogous to changing what was a slanging match to an orderly debate for which the neoclassical economists get to determine the language to be used and write the rules of order.

Conventional neoclassical economics is essentially concerned with biodiversity as a resource (Simmons 1993, 18–45; Heltne, in this volume) that, while renewable, is destructible (Randall 1986, 79–109). The maintenance of constant real consumption expenditure over time (maximum sustainable income) (Hicks 1946) requires the maintenance of the value of the asset base, which includes natural resources (Solow 1986). Whether within this paradigm renewal or destruction of biodiversity is a benefit or a loss becomes a province of calculation, that is, cost-benefit analysis (CBA). CBA weighs the costs of proceeding with a strategy as opposed to the benefits that could be derived from it. In order to carry out this procedure, monetary values are assigned to identifiable costs and benefits, both those that have commercial values found in the market and those considered in an "extended utilitarian accounting" that are found within the human preference structure but not exchanged in the market (Beder 1993; McNeely 1988, 1–36).

The philosophical premises of CBA are utilitarian, anthropocentric, and instrumentalist: "utilitarian in that things count to the extent that people want them; anthropocentric, in that humans are assigning the values; and instrumentalist, in that biota is regarded as an instrument for human satisfaction" (Randall 1988, 218). Two powerful epistemological assumptions lie at the heart of this microeconomic perspective. First, decisions concerning the use of natural resources are typically made by private individuals or corporations seeking to maximize their own welfare through the operation of competitive markets; and, second, only this process will ordinarily yield an allocation of resources approximating the social optimum. Given these assumptions: "Effective systems of management can ensure that biological resources not only survive, but in fact increase while they are being used, thus providing the foundation for sustainable development" (McNeely 1988, 2). Conservation becomes, in this instance, a means for sustainable economic growth, and we have what is necessary to construct independent provinces of both knowledge and ignorance-squared.

The flaw in the CBA approach, as is true for much of social and behavioral science, is that, when it comes to theories of complex phenomena, making them more analyzable may be convenient but does not necessarily make them better at describing what is actually occurring (Gell-Mann 1995, 323). The interdependency of species and ecosystems indicates that any calculations of losses from an extinction of a species in the present must take into account the loss of future benefits from other dependent species that also succumb. Therefore, in order to calculate the value of a particular species, the values of all other species would somehow have to be determined. In theory, ecological information could be factored into these calculations, "but of all the areas of biology and ecology, few are less understood than the inter-specific dependencies" (Norton 1988, 203).

The decision to base natural resource policy on the results of CBAs is rhetorical in its ignorance-squared selectivity. It is evident that valuation models based on economic calculation alone will pose ethical difficulties. There may be "good reasons to oppose efforts to put dollar values on non-marketed benefits and costs" (Kelman 1990, 129). The ultimate validity of such models depends on several assumptions. First, one must accept that economists are able to control for all dimensions of quality. Second, one must assume that the nonmarket entity affects all people equally. Third, one must assume that there is no difference between the price a person is willing to pay to get something and the price he or she is willing to pay to avoid giving up something. Fourth, one must assert that citizens do not differentiate between values expressed in private transactions and those expressed in public policy decisions (Peterson and Peterson 1993, 59).

However, before the economist even begins to construct the cost-benefit table, he or she is confronted by immense difficulties of methodological fuzziness, normally hidden away in the mathematical formalism that so enthralls policymakers. These difficulties can be grouped under four headings: (1) sustainability, (2) irreversibility, (3) "arrows of time and thermodynamics," and (4) complexity.

First, as Lele (1991, 613) argues: "[Sustainable development] is a 'metafix' that will unite everybody from the profit-minded industrialist and risk-minimizing subsistence farmer to the equity-seeking social worker, the pollution-concerned or wildlife-loving First Worlder, the growth-maximizing policy-maker, the goal-oriented bureaucrat and, therefore, the vote-counting politician." Economists have little difficulty conceptualizing sustainability. For instance, in general equilibrium economic welfare theory, *sustainability* would refer to "the maximum amount that could be spent on consumption in one period without reducing . . . real consumption expenditure in future periods" (Boadway and Bruce 1984, 9). The difficulty, as Bromley has noted, is that "the existence of a market still requires the willful coming together of two consenting agents to exchange for mutual gain" (1991, 87). Except for overlapping generations, this "willful coming together" is impossible across generations, providing, as discussed above, yet another example of a "missing market." This raises doubts about the ability of "market" analysis to achieve either intergenerational efficiency or equity, but it is neither a problem nor even an issue for those who purposefully choose not to know or care.

Second, there is now a considerable body of economic literature that deals in general with the problem of irreversible decisions in the face of uncertainty (this is largely dealing with the crux of one of the points elicited in Talbott, in this volume). The pioneering economics paper is that of Arrow and Fisher, who considered the problem of developing wilderness when the future benefits from conservation and development are uncertain. They show that, since development is irreversible while conservation is not, there is a value, which they called the "quasi-option value," associated with the reversible decision to conserve. Simply put, with uncertainty present, there is some value associated with keeping one's options open (Arrow and Fisher 1974). This is popularly known as the "precautionary principle" and is considered less than scientific, and, therefore, given little credence, by most economists. That is, the lack of "positive scientificity" relegates the precautionary principle to the province of purposeful ignorance by the economist.

Third, in the context of "arrows of time" and thermodynamics, it is not that orthodox economists are unaware of the relevance of thermodynamics in analyzing economy-environment interactions (Binswanger 1993, 209). It is simply that, for most economists, it remains an open question whether the second law of thermodynamics really matters (see Spreng 1984; Khalil 1990; or Lozada 1991). As long as economic systems predominantly used renewable resources and did not exploit them to exhaustion, entropy increases were never deemed a specific problem by economists, even though, as Wes Jackson (in this volume) points out with respect to agriculture, "over ten millennia, we have glided from one energy-rich carbon pool to another. First it was soil carbon, then forest carbon, then coal, then oil, then natural gas." The extensive use since the industrial revolution of nonrenewable resources has even more radically changed the impacts of economic systems on the environment. Today, economic systems mainly function outside ecocycles. Since nonrenewable resources such as coal, oil, and uranium are used in a way that is not adapted to the ecocycles, outputs of high entropy (pollution, solid waste, heat) cannot be recycled within the terrestrial ecosystems. This situation causes entropy increases in the environment, which leads to irreversible changes (deforestation, climate changes, extinction of species, etc.). So, while the industrial revolution has substantially advanced capitalism during the last 250 years, economists concerned with the resulting methodological implications of entropy have been relegated to the margins of the discipline.

Fourth, and finally, the complexity of the interaction between the economy and the environment limits the application of CBA. This type of analytic technique requires one to suggest that the problems are well defined, that the options are clearly specified, and that the political wherewithal is available to respond. All too often, the apparent objectivity of CBAs is the result of slapping arbitrary numbers on subjective judgments and then assigning the value of zero to the things that nobody knows how to evaluate. That is, the nonmarket value of viewing an elephant (or any other threatened species) in the wild is, to an economist, a function of how many people visit South Africa, how many of these people care whether they see an elephant in the savanna when they visit, and whether they would pay to see one again. The value that our children's children may put on the opportunity to view the elephant in the wild is assumed to be zero, and no compensation for the possibility of not being able to see one in the future is ever considered because it has no place in the logical construct of knowledge within which the economist works.

CONCLUSION

As F. A. Hayek once said: "Perhaps it is only natural that the circumstances which limit our factual knowledge and the limits which thereby result for the application of our theoretical knowledge are rather unnoticed in the exuberance, which has been brought about by the successful progress of science. However, it is high time that we took our ignorance more seriously" (1972, 33). The neoclassical belief that economics can reveal the certainties of the world by universalizing and essentializing concepts of self-interest, efficiency, and the exigencies of the competitive market has yet to be undermined by the collapse of certainty taking place within many of the physical sciences on which economics has been styled.

Economists, led by those presently at the margin of the discipline, must begin to emphasize bounded rationality and the dynamics of evolution and learning. Instead of basing theory on assumptions that are mathematically convenient, models must be constructed that are psychologically realistic. Instead of viewing the economy as some kind of Newtonian machine, it must be recognized as something organic, adaptive, surprising, and alive.

Specifically, to counteract ignorance-squared, economists on the margin must continuously attempt to identify a broader base of what can be known. Encourage students to search for alternative voices to the neoclassical texts, facilitating their search with selected readings and references to Marxist, institutional, or post-Keynesian analysis. Break the Anglo-Celtic dominance by providing students with economic exegesis and analysis of work that has been done and is being done in Japan, China, India, the Arab world, and Latin America. Encourage comparative examination and analysis of alternative streams of economic thought with respect to issues of poverty, equity, efficiency, production and distribution, etc. Identify the implications of ignoring nonmarket or limited-market economic activity of subsistence agricultural or pastoral communities in less economically developed countries. Insist on discussion of the political, cultural, and social ramifications of economic policy decisions and/or the determination of economic policy within a framework of power relations. Write for and communicate with others outside the discipline in the areas of the environment and ecology, sociological and political ramifications of economic power relations, or other less mathematically elegant fields of endeavor. Finally, continue to insist on the broadening of economic research and teaching within the discipline to incorporate multidisciplinary and multi-intellectual frameworks of analysis. In other words, recognize by all means the necessity of remaining humble in the face of ignorance, but struggle to break up the purposeful production of ignorance-squared.

Unfortunately, attempts at communication are largely ignored by neoclassical economists, and their lack of desire to engage in conversation with others is then passed on to the next generation in the classroom. The motivated tendency to tailor one's opinions in accordance with perceived audience preferences has been long recognized by social psychologists as a feature of the process whereby people, in general, form their judgments (Kruglanski 1991, 227). The perceived audience for most graduate students is their neoclassical mentors.

And so it continues. University departments, professional journals, and peers form an institutional web that provides for the career potential of any aspirant to the profession (North 1990, 95). The proficiency shown in neoclassical tools, concepts, and language becomes the hallmark of identification and quality. The Krueger Commission on Graduate Education, established in the United States to report on tertiary education standards in 1990, reported that department procedures "bias the selection towards good technicians, rather than good potential economists." This implies that graduate education deemphasizes creativity and problem solving as the student requires "little or no knowledge of economic problems and institutions" (Krueger 1991, 1040–42). Consequently, ignorance-squared is promoted as a qualitative manifestation of a "good economist."

The stated concern herein has been that neoclassical economists, as traditional intellectuals, cultivate the social production of ignorance-squared in the struggle for ideas. This is done through narrowing the pedagogy (mathematical modeling), confining research parameters (excluding social, cultural, and institutional analysis as noneconomic), and placing constraints on the production and presentation of alternatives to Anglo-Celtic and mainstream economic analysis. The worldview of the neoclassical economist continues to be based on a priori, logical, and mechanistic mental constructs, on deterministic processes that are always approaching equilibrium, on the presumption of social and cultural homogeneity in global markets, on the relegation of nonmarket activities to other disciplines, and on structurally simplistic concepts of supply and demand. And this generation of mainstream neoclassical economists will provide the next generation of economists with this limited worldview, having themselves been made purposefully ignorant of the evolutionary, complex, and ever-changing processes of a world permeated with the unintended consequences of intentional action rooted in ignorance-squared.

REFERENCES

Arrow, K. J., and A. C. Fisher. 1974. "Environmental Preservation, Uncertainty and Irreversibility." *Quarterly Journal of Economics* 88:312–19.

Arrow, K. J., and L. Hurwicz. 1958. "On the Stability of the Competitive Equilibrium." *Econometrica* 26:522–52.

Beder, S. 1993. *The Nature of Sustainable Development.* Newham, Victoria: Scribe.

Binswanger, M. 1993. "From Microscopic to Macroscopic Theories: Entropic Aspects of Ecological and Economic Processes." *Ecological Economics* 8, no. 3 (December): 209.

Boadway, R., and N. Bruce. 1984. *Welfare Economics.* Cambridge: Blackwell.

Bromley, D. W. 1991. *Environment and Economy: Property Rights and Public Policy.* Cambridge: Blackwell.

Caldwell, B. 1982. *Beyond Positivism: Economic Methodology in the Twentieth Century.* London: Allen & Unwin.

Chalmers, A. 1982. *What Is This Thing Called Science?* 2nd ed. St. Lucia: University of Queensland Press.

Coleman, L. 1998. "The Age of Inexpertise." *Quadrant* 42, no. 5 (May): 63–67.

Collingridge, D. 1980. *The Social Control of Technology.* Milton Keynes: Open University Press.

Drepper, F. R., and B. A. Mansson. 1993. "Intertemporal Valuation in an Unpredictable Environment." *Ecological Economics* 7, no. 1 (February): 43–67.

Gell-Mann, M. 1995. *The Quark and the Jaguar: Adventures in the Simple and the Complex.* London: Abacus.

Hausman, D. M. 1989. "Economic Methodology in a Nutshell." *Journal of Economic Perspectives* 3, no. 2 (Spring): 115–27.

Hayek, F. A. 1972. *Die Theorie komplexer Phanomene.* Tubingen: Mohr (Paul Siebeck).

Hicks, J. R. 1946. *Value and Capital.* Oxford: Oxford University Press.

Hutchison. T. W. 1938. *The Significance and Basic Postulates of Economic Theory.* London: Macmillan.

Kelman, S. 1990. "Cost-Benefit Analysis: An Ethical Critique." In *Readings in Risk,* ed. T. S. Glickman and M. Gough, 129–35. Washington, DC: Resources for the Future.

Kerwin, A. 1993. "None Too Solid: Medical Ignorance." *Knowledge: Creation, Diffusion, Utilization* 15, no. 2 (December): 166–85.

Khalil, E. 1990. "Entropy Law and Exhaustion of Natural Resources: Is Nicholas Georgescu-Roegen's Paradigm Defensible?" *Ecological Economics* 2:163–78.

Kirman, A. 1989. "The Intrinsic Limits to Modern Economic Theory: The Emperor Has No Clothes." *Economic Journal* 99:126–39.

Koppl, R. 1990. "Price Theory as Physics: The Cartesian Influence in Walras." *Methodus* 2, no. 1:17–28.

Krueger, A. O. 1991. "Report of the Commission on Graduate Education in Economics." *Journal of Economic Literature* 29, no. 3 (September): 1035–53.

Kruglanski, A. W. 1991. "Social Science-Based Understandings of Science: Reflections on Fuller." *Philosophy of the Social Sciences* 21, no. 2 (June): 223–31.

Lele, S. 1991. "Sustainable Development: a Critical Review." *World Development* 19, no. 6:607–21.

Lozada, G. A. 1991. "A Defense of Nicholas Georgescu-Roegen's Paradigm." *Ecological Economics* 3:157–60.

May, R. M. 1976. "Simple Mathematical Models with Very Complicated Dynamics." *Nature* 261:459–67.

Mayhew, A. 1996. "AEA Economics Journals." *Review of Heterodox Economics,* Winter, 1–2.

McNeely, J. A. 1988. *Economics and Biological Diversity: Developing and Using Economic Incentives to Conserve Biological Resources.* Gland: International Union for Conservation of Nature and Natural Resources.

Negishi, T. 1962. "The Stability of a Competitive Economy: A Survey Article." *Econometrica* 30:635–69.

Norgaard, R. B. 1989. "Three Dilemmas of Environmental Accounting." *Ecological Economics* 1:303–14.

North, D. C. 1990. *Institutions, Institutional Change and Economic Performance.* Cambridge: Cambridge University Press.

Norton, B. 1988. "Commodity, Amenity, and Morality: The Limits of Quantification in Valuing Biodiversity." In *Biodiversity,* ed. E. O. Wilson, 200–205. Washington, DC: National Academy Press.

Peterson, M. J., and T. R. Peterson. 1993. "A Rhetorical Critique of 'Non-Market' Economic Valuations for Natural Resources." *Environmental Values* 2, no. 1 (Spring): 47–65.

Randall, A. 1986. "Human Preferences, Economics, and the Preservation of Species." In *The Preservation of Species: The Value of Biological Diversity,* ed. Bryan Norton, 79–109. Princeton, NJ: Princeton University Press.

———. 1988. "What Mainstream Economists Have to Say about the Value of Biodiversity." In *Biodiversity,* ed. E. O. Wilson, 217–23. Washington, DC: National Academy Press.

Ravetz, J. R. 1993. "The Sin of Science." *Knowledge: Creation, Diffusion, Utilization* 15, no. 2 (December): 157–65.

Samuels, W. J. 1991. "Truth and Discourse in the Social Construction of Economic Reality: An Essay on the Relation of Knowledge to Socioeconomic Policy." *Journal of Post Keynesian Economics* 13, no. 4 (Summer): 512–24.

Sandven, T. 1995. "Intentional Action and Pure Causality: A Critical Discussion of Some Central Conceptual Distinctions in the Work of Jon Elster." *Philosophy of the Social Sciences* 25, no. 3 (September): 286–317.

Simmons, I. G. 1993. *Interpreting Nature: Cultural Constructions of the Environment.* New York: Routledge.

Simon, H. A. 1976. "From Substantive to Procedural Rationality." In *Method and Appraisal in Economics,* ed. S. Latsis, 129–48. Cambridge: Cambridge University Press.

Smith, A. 1910. *The Wealth of Nations.* London: Dent. Originally published in 1776.

Smithson, M. 1989. *Ignorance and Uncertainty: Emerging Paradigms.* New York: Springer.

———. 1993. "Ignorance and Science." *Knowledge: Creation, Diffusion, Utilization* 15, no. 2 (December): 133–56.

Solow, R. M. 1986. "On the Intertemporal Allocation of Natural Resources." *Scandinavian Journal of Economics* 88:141–49.

Spreng, D. 1984. "On the Entropy of Economic Systems." In *Synergetics—from Microscopic to Macroscopic Order,* ed. E. Frehland, 207–18. Berlin: Springer.

Tversky, A., and D. Kahneman. 1987. "Rational Choice and the Framing of Decisions." In *Rational Choice,* ed. R. Hogarth and M. Reder, 67–94. Chicago: University of Chicago Press.

EDUCATING FOR IGNORANCE

Jon Jensen

Is it possible that education is a process of trading awareness for things of lesser worth? The goose who trades his is soon a pile of feathers.
—Aldo Leopold

What is education for? On the surface, this is an odd question because few of us doubt that we know the goal or desired outcome of education. We may debate the particulars of a curriculum or whether a given school is succeeding or failing, but don't all of us agree that knowledge is the goal of education? What would be the point of schooling if not to "learn stuff"? Even the often-repeated aphorism that education is about "lighting fires, not filling buckets," is generally justified by the observation that students learn more, that is, acquire more knowledge, when they are motivated to learn by dynamic teachers. Those special individuals who light fires in their students are, ultimately, more successful at filling buckets, upping the quotient of knowledge that students absorb and retain.

I wish to explore the possibility that even this most basic assumption—that the primary goal of education is the acquisition of knowledge—might be flawed. What if education is, ultimately, about ignorance more than about knowledge? What might we learn about our schools, and ourselves, by focusing on ignorance and human limitations, rather than knowledge, in thinking about what and how we teach? I don't mean to draw attention to ignorance as the starting point for education since seeing ignorance as the problem to be solved is perfectly compatible with educating toward the goal of maximizing knowledge. Rather, I want to suggest, or at least explore the possibility, that ignorance of a certain sort might be a proper end point, one goal, for a successful educational system, or at the very least a useful tool in thinking about education.

It is not ignorance per se that is my concern but ignorance of a certain sort and, more precisely, a perspective that embraces the inevitability of human ignorance, rather than seeing it as a problem to be cured. Let me be clear. The goal is *not* to increase the amount of ignorance—we have an excess of that already—but to instill an awareness of the limitations of human knowledge and power. What this would require is a shift in focus to recognize that the basic purpose of education is not to manufacture consumers and workers but to nurture citizens of communities—both human and natural. Students must gain a perspective, a way of seeing the world and their place in it. This goal will certainly be criticized for being vague and amorphous and for not lending itself easily to specific definition or application. But it is no more so than justice or other ideals that guide our thinking and action.

To achieve such a perspective, educational reform might begin not with discussions about whether testing should take place during the third or fourth grade, or whether one or two science courses should be part of the required curriculum, but with the basic and fundamental questions: What is education for? How does education prepare people for the work of the world within a culture and a place? What do we hope for in our graduates? What would happen if we put ignorance, rather than knowledge, at the center of our thinking about education? How might it affect our educational institutions to strive for graduates who appreciate their own ignorance, rather than basking in their knowledge?

This notion of educating for ignorance is, admittedly, a counterintuitive idea and one that is vulnerable to the dangerous misinterpretation that we should not value education, should not strive to gain more knowledge. But I wonder if it isn't equally ripe in its prospects for helping us rethink education in a way that is essential to the health of our culture and the biosphere.

Before saying more about changes in education, it's necessary to flesh out a bit my use of the term *ignorance* and particularly this idea of a perspective or worldview centrally focused on ignorance. For me, this is best done with the story of two wooden objects juxtaposed on my small farm and the contrary perspectives they embody.

As I look out my dining room window past our deck to the woods beyond, I see a mighty bur oak tree. This tree was standing when Norwegian immigrants settled here in the 1850s, and it serves as a land-mark, a sort of standard by which to measure other, newer elements of this small farm. This bur oak, a remnant of the prairie, has lately seemed to stand in judgment of the deck attached to the west side of the house. Even though I've spent countless hours eating, drinking, talking, and reading on that simple wooden platform, these days it feels less like a familiar and cherished place than a sign of human recklessness and shortsightedness.

It's a typical deck with planks and railing weathered gray in that familiar color of aging wood. In this case, as in many others, the familiarity masks its toxicity. Pressure-treated lumber is common throughout the United States, not only in decks, but also in playground equipment, backyard swing sets, and most outdoor uses. The label *pressure-treated* is really something of a misnomer, however, since it sounds like a simple mechanical process rather than infusion with a chemical cocktail. Chromated copper arsenate (CCA) is the chemical mixture used to prevent rot in the wood, but it's the 22 percent pure arsenic—a known human carcinogen—that is the concern. The arsenic in pressure-treated wood rubs off on the hands of people who touch it, which is why the manufacture of CCA lumber for residential use was banned as of 1 January 2004.

I almost certainly have been exposed to arsenic, as have most Americans. It's one of the hundreds of industrial chemicals that all of us encounter every day. The arsenic may or may not hurt me, but studies show that children exposed to arsenic have significantly higher incidences of bladder and lung cancer. I don't know whether anyone will become sick from the chemicals in my deck. Nor do any of us know who will be the victims of exposure to the tens of thousands of other chemicals now common in our society.

It is this "not knowing" that is the key element of the story for me, not just with pressure-treated wood or the chemicals. In many, if not most, cases, we simply do not know the effects of our actions, especially when dealing with technologies and human inventions. Yet we nearly always proceed on blind faith, assuming that, since we humans were smart enough to create the situation, then we will be smart enough to deal with it. We invent new and better devices to solve all our problems with little thought for what additional problems might be created or who might be harmed by this "advance." Whether it's nuclear waste, genetically modified foods, leaded gasoline, thalidomide, or any of countless other developments, the story is the same. We shoot first and ask questions later. We are technological lemmings, jumping off the cliff without sufficient thought about what lies ahead. What drives this process is a boundless confidence in our ability to solve all problems, to design new and better devices to patch up any unintended consequences of the previous devices. It's not that the designers and engineers don't think about consequences or test their products, though clearly more and longer-term testing is called for. Rather, it's the attitude behind the scenes, the assumption that we can handle it, that we know enough, are powerful enough, and quick enough to make it all work out.

The example of pressure-treated wood helps illustrate the basic idea and the motivating force behind an ignorance-based worldview. It is primarily a shift in perspective, a change in assumptions that would then result in new approaches and, thus, different actions. This shift is motivated by a study of history and a recognition of the failed assumptions that have caused so much destruction of people, land, and communities. Especially in the last half of the twentieth century, and especially in the United States, humans have assumed that we know enough to manage the world and alter it to serve our short-term purposes. This seemingly simple assumption about our knowledge—that we *know enough*—cuts to the heart of the difference in perspective between the bur oak in my yard and the humans who created the CCA used to treat my deck. It's the difference between nature's wisdom and human cleverness, between the ecological and the anthropocentric. The embedded knowledge of the natural world is boundless and beyond our reach, except in little glimpses, most of which we ignore. In the Industrial Age, we assume that human knowledge is even greater than nature's wisdom, and we proceed as if nature was merely the inert backdrop for the play of human dominance. The alternative, ecological perspective takes human ignorance as a given, the central fact of our existence, the context that must never be forgotten.

Focusing on ignorance challenges the assumption of human knowledge, suggesting that, in fact, humans know very little, especially about the workings of the natural world. We are, in Wes Jackson's words, "billions of times more ignorant than knowledgeable," and, more important, this condition is neither temporary nor curable. Our partial and incomplete knowledge, our ignorance, is an inevitable part of the human condition. Recognizing this fundamental, essential, and incurable ignorance would radically alter the way in which we act and halt many destructive processes.

Some may argue that I overstate the point, that the case of pressure-treated wood simply shows the need for more testing and more care in the development of new chemicals and novel technological applications. Our technological capability has evolved rapidly, and we have struggled to keep up on the regulatory front. But is this cause for a new worldview? Is it truly a sign that we need to fundamentally change our society and its educational system? The ban on arsenic in treated lumber is seen by some as a clear indication that the basic system is working; we discovered the problem and took steps to fix it. More caution (and corresponding regulation) is what we need, not some new notion of human ignorance.

This response assumes that the problem is small and requires only a slight tweaking of the system: a few new laws, a little more testing, slowing down the process a bit. But treated wood, chemicals, and technology are only the tip of the iceberg, the iceberg being an attitude that pervades our society. Sufficient knowledge and human superiority are embedded in our buildings, our business and social structures, and the fabric of our culture in a way that additional legislation cannot unseat. No tweaking of chemical testing will, ultimately, change this basic attitude or alter its effects. William McDonough and Michael Braungart make a similar point in *Cradle to Cradle,* their book on ecological design, with a provocative chapter called "Why Being Less Bad Is No Good."[1] The basic point—if something is fundamentally flawed, then making it slightly less bad is no significant gain—gets to the heart of my point here. Arsenic is merely a symptom, a downstream effect of a deeply engrained way of thinking that advances on the basis of what we know with little thought for limits, ecological context, or all that we don't know about consequences and connections.

The objection is right in the sense that all this talk of ignorance is really not new. Recognizing the limits of human knowledge is, in fact, a very old idea, literally an ancient attitude. Indigenous peoples have guarded against the abuses of knowledge in their religions and cultural practices for millennia, and hubris was a vice that plagued the Hellenic world no less than it does ours. What is new is not the existence of the hubris that attaches to progress, to a blind devotion to the new, but the control that this attitude has over our contemporary society and the potential for destruction that comes with its marriage to technology.

When the Greeks believed that their knowledge was sufficient to run the world, they wrought havoc on a limited geography. When we assume that our knowledge is sufficient, we risk the irremediable extinction of species, the mass destruction of life through nuclear contamination, and a loss of soil that could threaten the possibility of food production for millennia. The attitude is the same, but the costs are higher.

Acknowledging the inevitability of human ignorance could transform more than chemical manufacturing. Our educational system would inevitably be different as well. If we admit that students can never *know enough,* then we must go beyond asking what is most important to know to inquire what sort of attitude students should have about what they know. What perspective on human knowledge and our place in the world is reflective of a well-educated person?

The aim of any school or educational system should be to graduate students with a certain character. Lists of objectives, statements of intent, talk of worldviews, all this means little. The success of our educational system should be judged on the basis not of what graduates know but of their attitude toward what they know. It's about character, not knowledge, about the virtues that guide individual lives. In thinking about what it would mean to educate for ignorance, I suggest the following six character traits as examples of what we need to instill in graduates of our schools, colleges, and universities: propriety, interdependence, humility, respect, conservatism, and awareness.

Propriety is an old-fashioned notion, but one that has never been more important than for today's graduates, who must recognize the fit and appropriateness of themselves to the land, our individuality within community, and our actions within their context. To be guided by propriety is to be ever conscious of connections, to ask: What is appropriate here? What will the land support? What will be the impacts of my actions on others, on community, on place? The opposite of propriety is the universality that says one size fits all and pays no attention to the particular demands of context and place. At a fundamental level, all this boils down to a simple value—restraint. Are we willing and able to restrain ourselves, especially in this time of rapid technological development? If humans know enough to solve every problem, then restraint and propriety seem like quaint elements of the past. If, however, we see human ignorance as our inescapable condition, then setting limits and finding our place become the fundamental work of humans in the world.

Graduates who acknowledge human ignorance will recognize connections and reject the atomization, specialization, and reductionism that are typical of much contemporary thinking, especially in the sciences. Thinking in terms of systems and interdependence is a habit of thought rooted in the fundamental law of ecology that "all things are connected." The arsenic in my deck does not affect only me; it ripples throughout the system and has impacts that are far beyond my comprehension. To recognize that you cannot do just one thing is essential to appreciating one's ignorance and, thus, to reining in one's actions. With such interdependence comes the recognition that nature, not humanity, is the proper measure for our actions, our thought, and our decisions. This is not to suggest that humans are separate from nature or that nature is separable from us. It is, however, to recognize that all human actions, from the work of planting a garden to the writing of computer code for the Mars rover, take place within the context of the biotic community in which we live. The only known measures we have for longevity and for health are functioning ecosystems. To take man as the measure, as we so often do, is to live in the arrogance and shortsightedness that characterize our time.

Nothing comes closer to the heart of an ignorance-based worldview than the virtue of humility. The notion that we must reject the hubris and arrogance of our time and embrace an attitude of humility, a recognition of our limitations and failings, is absolutely crucial to beginning to shift our culture in a new direction. The humility that comes from recognizing our fundamental ignorance manifests itself in respect for ourselves, other living things, communities, and the land. Respect is an attitude that is difficult to define precisely, but overcoming our culture's tendency to control and dominate all things is key to this shift in perspective. Such disrespect is everywhere in our society, from obesity to soil erosion, from Wal-Mart to genetically modified organisms; the examples abound. Appreciating ignorance calls for shifts in thought and practice to focus on humility and respect as prime virtues.

Another way of capturing this attitude of humility is by saying that we need graduates who are more conservative. Of course, I do not mean that we need graduates who are more likely to vote Republican. The word *conservative* must be reclaimed from the political Right, whose policies generally conserve little besides the wealth of the party faithful. To be genuinely conservative, to go slowly and not automatically commit oneself to the newest trends, to avoid unnecessary chances without a clear sense of the repercussions, is a trait central to the humble individual who respects natural systems. Conservatism is the opposite of the blind drive toward the future, toward progress and perpetual growth, a view that dominates both the Right and the Left in contemporary politics. But a blind acceptance of all traditions is similarly antithetical to the spirit of an ignorance-based worldview. An absolute resistance to change can be based only on the assumption that we have it all figured out and that we *know enough* to reject the possibility of a better way of doing things. Conservatism is not opposition to change but a caution that necessitates slow and thoughtful change. We need, paradoxically, graduates with a sort of progressive conservatism that moves us beyond the status quo while seeing the big picture and carefully studying exits and indirect effects.

Finally, awareness is a fundamental trait that must be nurtured in all students. Each of us is born with an innate curiosity and a desire to see and experience all that is around us. This awareness or fascination with our surroundings is the source of the wide-eyed infant, hyperaware of every sight, sound, and smell. But, as Aldo Leopold noted over fifty years ago, we tend to lose this awareness as we grow up, often trading it for the obliviousness and self-absorption that we call sophistication. As both individuals and a culture, we have adopted the adolescent tendency to see the world as revolving around us with very minimal awareness of our place in the whole or the constant feedback we are receiving from the "Other." Regaining a heightened awareness, an ability to listen with all our senses to the voices, human and other, that compose the living world, is a crucial component of adopting an ignorance-based worldview. Acknowledging our ignorance makes such listening a necessity since we are put in a constant position of needing to know, needing to be aware.

Striving to instill these virtues helps focus our discussion of ignorance as well as providing some direction for thinking about educational reform. But one hopes that it also helps avoid one crucial point: ignorance itself is not a virtue. Ignorance is not something to strive for; rather, it is an inescapable part of being human. Recognizing this human condition leads not to a passive complacency but to an active state of awareness and learning. The actively ignorant person is trying not to gain enough information, because she knows that this is impossible, but to be constantly aware of the implications of her decisions and to always think about, and leave room for, second chances.

What is necessary to move education in the direction I am suggesting, toward graduates who appreciate human limits and possess some of these virtues? Of course, no silver bullets exist, and educators have become wary of the next educational reform proposal. What I wish to suggest is not some new system but, rather, some shifts in how we teach and what we teach. For example, I've been arguing that we should put less emphasis on content and more emphasis on character. This does not mean teaching less content; rather, it means seeing content as a tool, not an end. While it's crucial that we not dumb down our educational system, it's more important that we develop and nurture a certain type of person. Rather than simply producing graduates who have acquired a body of knowledge, as if it were one more commodity to be consumed, we need graduates with a perspective and associated character traits. Obviously, there can be no character development without significant content, but the converse is not true—one can, and many people do, learn countless bits of information with no impact on their character.

Nowhere is this more obvious than in the tendency toward mastery of a specific body of information for a test. Such a focus on mastery masks what is not known—including what can never be known—while narrowing a student's focus. Education for ignorance may teach the same material but in a way that expands awareness rather than narrowing, thus exposing all that is unknown and creating a very different perspective in the student. Whether it's schoolyard gardens, interviews with local elders, or science experiments, developing an appreciation for human limitations is more likely to happen in ways that go beyond simple memorization and mastery of content. The growing interest in inquiry learning, especially in the sciences, is an encouraging trend here since it teaches students about the nature of science—and, thus, inevitably, its limitations—rather than giving the idea that science is simply "the truth." The basic idea here is to focus science education on questions rather than answers, encouraging students to inquire about the subject matter rather than learn the facts.

We also must shift the emphasis away from machines and technology and toward living organisms and systems. Regaining the study of natural history, which has almost completely faded from curricula at all levels, would be a good first step. When students learn about and connect with their place, this increases awareness and helps nurture interdependence and respect. Putting computers and technology in proper perspective is a major hurdle in this area. We must recognize that technology is not an educational end in itself and cease to treat it as such. While computers are helpful, they are tools, and we must teach students to evaluate the impact of technologies on our lives and our communities. Here, as in many areas, we have much to learn from the Amish, not so much in their choices of technology as in the way in which they measure new developments against the standard of community.

Perhaps no change is more important than reemphasizing place in education at all levels. Developing a connection to the local environment and a sense of place is crucial to appreciating human ignorance and the humility that accompanies it. The movement toward more place-based education is encouraging, though still very small. Here in Iowa, some schools are restoring portions of school grounds to tallgrass prairie and educating students about the native systems obliterated by industrial agriculture. These small steps are encouraging, but, still, the average Iowan knows more about the tropical rain forest or the ocean than about the seas of grass that once covered our state. Educators need to recognize the value of studying local history and culture and resist the temptation to limit ourselves to the universal content of standardized tests.

There are many other changes in emphasis that must be made, but the key is starting the process of reexamination and recognizing that no one solution will be universally applicable. The changes will need to evolve, finding what fits in a given place, at a particular time. In addition to these shifts in how we teach and what we teach, a paradigm shift in education will also necessarily affect the basic structure of schools, both physically and organizationally. Our schools are like prisons, holding and controlling the inmates we call students and constructing strict hierarchies of power and decisionmaking.

No change in school structure is more important than addressing issues of scale. Most schools today are behemoths, often with thousands of students, completely oblivious to the notion of appropriate scale that is essential to humility. Both classes and schools must be smaller in order not only to facilitate many of the changes outlined above but also to teach students about propriety and the importance of considering scale in all aspects of life. Smaller scale would help facilitate organizational changes as well, moving from a top-down, hierarchical structure to one that mimics natural processes through organicism and interdependence.

Changes in the physical structures of schools are also vitally important. The movement toward edible schoolyards gets students outside with their hands in the dirt, learning what they simply couldn't within concrete block walls. We must transform both campuses and classrooms. We need buildings that teach, and we need to teach outside buildings. We must eliminate the separation between education and work so that students spend time producing food and helping meet their other necessities as part of everyday life. Ecological design is leading the way in creating working examples of structural changes in schools.

In addition to these shifts in emphasis, we need working models of new educational approaches and guiding concepts to help steer reform. I have been intentionally vague about different levels of education up to now because I think that many of the same questions are important from the preschool through the doctoral level and beyond. But, obviously, there are differences as well. For primary and secondary students, the interrelated concepts of inquiry and place can come together to guide all curriculum and pedagogical decisions. The work of David Sobel is pioneering in mapping out what this would mean for teachers and schools.[2] It's exciting to think what would happen if students were outside exploring their local ecosystem every day and were encouraged not just to learn facts but also to ask questions as they discover. Specifically, the learning model called "Environment as an Integrating Context" is encouraging. Here, rather than disciplinary separation of reading, writing, math, and other subjects, all parts of the curriculum are linked to the environment in a way that encourages individual exploration. The focus is on local knowledge rather than universal concepts, and the model emphasizes making connections instead of memorizing facts. While this clearly changes the content of elementary education, it also influences attitudes and nurtures the development of character traits such as humility, respect, and awareness.

With higher education, I suggest that the concepts of sustainability and ecological citizenship may provide guiding principles for thinking about reform. How do we rethink university curricula so that every student understands his or her impact on the world and every major directly addresses issues of sustainability? How do we instill a sense of citizenship in the biotic community in all graduates? Many individual efforts are under way to answer these questions, with much effort currently focused on climate change. As always, the danger is a too-narrow, technical understanding of sustainability that leaves core assumptions untouched. Every first-year student in college should engage in analyses of campus systems and resource flows of energy, food, water, waste, and materials. No one should graduate without a basic understanding of the ecosystem in which he or she has lived and studied. David Orr's work on education provides a great resource, as does the work of a small group of innovative colleges called the Eco League.[3] The point here is not simply to understand more about sustainability but to recognize how little we know, and is known by anyone, about the basic biological systems on which we depend.

Some will object that none of this is possible: that it would be too expensive, that the teachers are not willing, that the students won't do it, that the parents will revolt. I cannot say that I am optimistic that major changes in education are right around the corner. But, even if I'm not optimistic, I am definitely hopeful. Change will come because it must. Neither our communities nor the biosphere can survive the continuing onslaught that is perpetuated and led by the graduates of today.

On summer evenings, I like to sit on my deck, look west through the woods, and watch the last color fade from the sky. My thoughts often drift forward, wondering where our society is heading and what will be the consequences in an uncertain future for the people and places that I care most about. Lately, I have taken to speculating on the year 2045, the point at which my infant daughter will reach my current age. What will the world be like? How will it be different from today? What sorts of challenges will humans face, and how will they work to address them? Population growth. Climate change. Loss of biodiversity. Soil erosion. Deforestation. It's easy to be overwhelmed when I extrapolate current trends and contemplate likely changes, but for some reason I remain hopeful.

Sometimes I picture my daughter, Sylvia, forty years old in 2045, sitting outside as I am, with children playing nearby. Whenever I picture such a scene, the images are always fuzzy in details, but the outline is the same. It's a peaceful and happy scene that does not match the trends I see today for reasons I do not completely understand. The beauty of daydreaming about the future is that we are freed from having to picture all the details, especially of the transition that will take us from here to there, details that can bog us down and keep us from believing that change is possible. The scene I am picturing is idyllic, but not romantically so. I am picturing a world where human attitudes toward nature and each other, toward what we know and what we can never know, are quite different than the attitudes that dominate our culture. It is a picture that is simultaneously radical and mundane, featuring changes not only in how we live but also in how we think. Part of what makes this seem possible is that many, many people today already possess similar attitudes and have received an education that fosters this. Change is slow, but it's happening in ways that will grow as time and other pressures dictate.

Near the end of *A Sand County Almanac,* Aldo Leopold remarks that he has presented the land ethic as an evolutionary development since "nothing so important as an ethic is ever 'written.'"[4] Leopold's wisdom applies equally well to the idea of educating for ignorance. One can no more write a new educational scheme than one can write an ethic or a worldview. Thus, I have not proposed a plan or a model and certainly nothing like a curriculum. Rather, my comments are merely an initial foray into the territory of ignorance and what it would mean to educate for students who appreciate human limits. The necessary changes in education are both small and unimaginably large, simultaneously subtle and revolutionary. These seemingly insignificant shifts in emphasis might just result in an entirely new way of thinking about and conducting education.

The key is that all of us must engage in the conversation that begins by asking the big questions. What is education for? Is it simply a way of preparing more cogs for the machine of industrial capitalism? If so, then our crisis of education and the environment, of community and culture, will deepen, wreaking unimaginable harm on the systems of the earth. But, if we are capable of seeing a different answer, then a different future is possible. Recognizing the fundamental and incurable ignorance of *Homo sapiens* may provide one important step in the journey to an alternative future.

NOTES

1. William McDonough and Michael Braungart, *Cradle to Cradle: Remaking the Way We Make Things* (New York: North Point, 2002).

2. See David Sobel, *Mapmaking with Children: Sense of Place Education for the Elementary Years* (Portsmouth, NH: Heinemann, 1998), *Beyond Echophobia: Reclaiming the Heart in Nature Education,* Nature Literacy Series (Great Barrington, MA: Orion Society, 1999), and *Place-Based Education: Connecting Classrooms and Communities,* Nature Literacy Series (Great Barrington, MA: Orion Society, 2004). To learn more about environment as an integrating context, see http://www.seer.org.

3. See, e.g., David Orr, *Earth in Mind* (1994), 10th anniversary ed. (Washington, DC: Island, 2004), and *Design on the Edge: The Making of a High Performance Building* (Cambridge, MA: MIT Press, 2006). The Eco League is a consortium of colleges with a specific focus on the environment and a commitment to innovative teaching in environmental studies (see http://www. ecoleague.org).

4. Aldo Leopold, *A Sand County Almanac and Sketches Here and There* (Oxford: Oxford University Press, 1949), 235.

CLIMATE CHANGE AND THE LIMITS OF KNOWLEDGE

Joe Marocco

> The highest that we can attain to is not Knowledge, but Sympathy with Intelligence. I do not know that this higher knowledge amounts to anything more definite than a novel and grand surprise on a sudden revelation of the insufficiency of all that we called Knowledge before,—a discovery that there are more things in heaven and earth than are dreamed of in our philosophy. . . . Man cannot know in any higher sense than this.
>
> —Henry David Thoreau

QUESTIONING OPTIMISM

We live in a time of a great, deep-seated optimism. With the explosion of empirical knowledge since the birth of modern science has come a widely held confidence in our unquestionable ability to transcend ignorance. What science cannot tell us today it surely will be able to tomorrow. All that stands between us and the truth is time and research money. In short, we ardently believe the secrets of nature are well within our grasp. Our dogmatic faith in the limitless capabilities of scientific inquiry is so ingrained that it has become a fundamental part of our worldview, silently, yet powerfully, influencing the way we perceive and respond to adversity.

There are good reasons to question our optimism about scientific knowledge, however, especially when it is applied to complex environmental problems. The methods of science are effective only to the degree that they are successfully reductive; the interaction between human culture and nature defies reductionism; that is, it cannot be translated into a series of mathematical formulas or physical laws, the mainstay of traditional scientific inquiry. Reductionism tends to overlook the need to place environmental problems in a larger context, treating them as though they can be solved with advanced scientific knowledge, and ignoring the ethical and behavioral factors that lie at their roots.

This century finds us facing an escalating crisis of the natural systems on which we depend for the sustenance of life itself. With each new environmental difficulty we face, we continue to look first to science to prove its existence and provide an understanding of it, then to the purveyors of technology and policy to put in place measures to stem it. This style of environmentalism has its roots in our scientific and technological optimism, and it has appeared to serve us well enough in the past. However, the far-reaching problems of the twenty-first century are proving resistant to the traditional methods of environmentalism. One of these in particular, global climate change, is a serious challenge to our optimism about the power of science and technology and demands a new understanding of the limits of knowledge.

THE ORIGINS OF A WORLDVIEW: FRANCIS BACON AND "THE GREAT INSTAURATION"

The conventional optimism concerning the capabilities of science descends from a prevailing knowledge-based worldview.[1] Central to this worldview is a proverb: "Knowledge equals power." Ingrained in the Western mind is the belief that more knowledge will yield more control over problems and phenomena, regardless of magnitude or intricacy. This maxim is a cornerstone of our pedagogy, tying the way we teach, learn, and live to the notion that more and better knowledge will allow us to prevail over adversities of all kinds, current and future. We have great hope that science will "get us there someday," and we will reap the benefits, whether it is a cure for cancer, an unlocking of the mysteries of human genetics, or a discovery of a source of energy that is inextinguishable, "green," and able to meet the ever-growing demands of modern society.

The primacy of the knowledge-based worldview stems from both the putative infallibility of the methods science uses in its acquisition of knowledge and a rich history of successful scientific inquiries. Our optimism about science's ability to solve problems is rarely questioned, our trust in the sole sufficiency of scientific knowledge rarely shaken, even in the face of extreme complexity and uncertainty. While this rather unchecked confidence is often merited, there is very good reason to question just how far scientific knowledge can take us on its own and just how much faith we should have in our scientific endeavors.

Climate change presents us with an unusual combination of extraordinary complexity and the potential for far-reaching, dire consequences—characteristics that, I believe, give us ample reason to carefully question the basis for our confidence in scientific knowledge. Before we can do this, we must understand how our hubristic beliefs about the power of knowledge came into being; in this regard, we cannot overlook the writings of Francis Bacon. The importance of realizing the profound influence that Bacon has exerted over our thinking about knowledge, power, and progress cannot be overstated. The precepts of a four-hundred-year-old worldview can be found therein.

In 1620, the English statesman and scholar Bacon published the *Novum Organum,* which translates to "new instrument." The impetus for Bacon's work was to bring about the end of the "vicious authority" of the Aristotelian systems of knowledge and to replace those systems with "the real order of experience," that is, the method of inductive reasoning. While Bacon believed that the methods and philosophies handed down by the "ancients" were useful for "the affairs of civil life," he did not find them particularly enlightening in matters of the natural sciences, matters that he held in the highest regard. He believed the deductive methods of the Sophists to be the "anticipation of nature"; his new method, employing induction, was, instead, the "interpretation of nature."[2] His refusal to accept as perfect truth that which was handed down from antiquity prompted Bacon to define a method that arrives at the truth in a controlled and methodical way through the gradual and linear acquisition of knowledge. This new method would erase subjectivity and guarantee absolute truth through experiment and reason. Using it, man could, he believed, hold all the secrets of the universe within a very short time. In fact, he was certain that this new method would bring humankind nothing less than a complete and flawless understanding of nature.

Bacon dubbed this epistemological renewal "the Great Instauration," believing that "our only remaining hope and salvation is to begin the whole labour of the mind again," making man's quest from thence forward "to penetrate the more secret and remote parts of nature, in order to abstract both notions and axioms from things, by a more certain and guarded method." He would do this by putting forth a method that rejected anything that could not be proved with a combination of empiricism and rationalism. According to Bacon, at no point was the seeker of truth to take liberties and make leaps; knowledge should be built linearly, and one ought not to "jump and fly from particulars to remote and most general axioms."[3]

If this closely controlled method were used, Bacon believed, nothing could or should remain unknown or mysterious: "For that which is deserving of existence is deserving of knowledge, the image of existence."[4] According to Bacon, the human capacity for knowledge is limited only by a lack of tenacity and dedication to the method that he proposed. This faith in the limitlessness of our ability to know has translated into the current-day faith in progress, not only of science, but also of all human endeavors in general. Bacon's method promised to yield an unquestionable ability to amend and append to our knowledge of the natural world.

When attempting to replace the accepted and exclusive Sophism of the ancient Greeks, Bacon, ironically, created a new dogma, one that assumes scientific progress to be an absolute. It is this assumption, and the ensuing *expectation,* that is most applicable to the historical development of a knowledge-based worldview. Before Bacon, one acquired knowledge for the purposes of attaining wisdom; in the *Novum Organum,* Bacon put forth a very different conception of what knowledge is and how it should be used: "Only let mankind regain their rights over nature, assigned to them by the gift of God, and obtain that power whose exercise will be governed by right reason and true religion."[5]

Under aphorism 3 in the *Novum Organum,* Bacon penned a version of his most famous dictum: "Knowledge and human power are synonymous, since the ignorance of the cause frustrates the effect. For nature is only subdued by submission, and that which in contemplative philosophy corresponds with the cause, in practical sciences becomes the rule."[6] For Bacon, knowledge and power are one; the assumption implicit here is that power over nature is one of the desirable ends—and, perhaps, ultimately the only desirable end—for humankind. When read from the perspective of the twenty-first century, it seems that we have faithfully heeded Bacon's call. Our scientific knowledge and resultant technologies have placed us in an unprecedented position of power over nature, forcing changes on the natural world that could not have possibly been foreseen by Bacon and his contemporaries of the scientific revolution.

After Bacon, knowledge became a means to an end, elevating science to a position of respected and revered problem solver. The Sophists' idea of the solitary pursuit of wisdom was replaced by a vigorous, and sometimes destructive, application of knowledge to society's most difficult and pressing problems. In his essay "Bacon's Forms and the Maker's Knowledge Tradition," Antonio Perez-Ramos explains: "The token value and the regulative function of such concepts as scientific progress identified with a betterment of man's moral condition, has continued unabated, despite the glaring contradictions revealed by the war industry and by ecological perplexities."[7] Brian Vickers, writing in his essay "Bacon and Rhetoric," states this point more clearly: "Not only in the natural sciences but in every field of knowledge today we confidently expect that new research will expand our understanding of a subject, enlarge or redefine the framework or conceptual categories within which we see or think it."[8]

In addition to the expectation of progress, under the knowledge-based worldview there exists a dogmatic belief in the ability of science to amend itself, to fix the problems that result from the inappropriate application of scientific knowledge. This assumed self-regulating and self-correcting ability is entrenched in the scientific and public mind alike. The following statement made in 1934 by John Dewey embodies this belief: "The wounds made by applications of science can be healed only by a further extension of [the] application of knowledge and intelligence. . . . This is the supreme obligation of intellectual activity at the present time."[9] It is safe to say that, in that regard, not much has changed since 1934 and, furthermore, that the influence of Bacon's optimism concerning the scientific endeavor is without question today.

"Truth and utility are here perfectly identical"—this is Bacon's response to those who might question whether he has "established a correct, or the best goal or aim of the sciences" in the *Novum Organum*.[10] From this synonymity, we gather that, in seeking the truth, we necessarily find the best means by which to utilize it. In other words, Bacon saw no reason to doubt that humankind would use scientific knowledge correctly because, when one comes to the truth by inductive methods, the "utility" of that knowledge becomes exceedingly clear.

If Bacon is right, it should follow that, the more evidence scientists gather in favor of anthropocentric climate change, the more likely we will be to reform our causative economic, technological, ethical, and social systems. The trouble with Bacon's assumption is that, like the method he proposes, it is reductive and narrow. We do not always and necessarily act from truth, especially when doing so requires sacrifice, upheaval, or a perceived loss of personal freedom or power. Contrary to Bacon's optimism are the agents of self-promotion and self-preservation, both of which often operate outside the realm of knowledge and rationality.

The "Summary for Policymakers" of the 2007 report of the Intergovernmental Panel on Climate Change states, with "very high confidence," that "the globally averaged net effect of human activities since 1750 has been one of warming."[11] With this knowledge in hand, can we expect to be driving hydrogen-fueled vehicles in the next decade? Will a majority of us be powering our homes with solar panels, choosing carpooling over solo driving, turning our thermostats down in the winter and up in the summer? Not likely. To wit, scientists have known about anthropocentric climate change for the last twenty years, but America's oil consumption has grown substantially within that time period.

Better knowledge is only a small part of change; it is clear that other factors trump its power in driving decisionmaking, actions, and activities. This points to a definite need to supplement our knowledge-based worldview with a perspective that respects the ease with which science can be so easily ignored, not only by those in powerful government positions, but by the common person as well. In short, Bacon was wrong: knowledge is not synonymous with power or utility. Even if we can leave off the antiquated notion of our ability to predict and control nature, we must also cease to believe that knowledge, and knowledge alone, can change minds.

KNOWLEDGE-BASED ENVIRONMENTALISM AND CLIMATE CHANGE

Bacon's influence pervades all areas of Western thought, including mainstream environmentalism. Environmentalists utilize a knowledge-based method of problem solving, consisting, in the broadest terms, of the following steps:

1. Identify effects (biodiversity loss, climate change, air/water pollution).
2. Determine the causes (loss of habitat, greenhouse gas emissions, etc.).
3. Address the causes through a variety of measures (promotion

of "green" technology, government lobbying, ethical/moral/aesthetic appeals to the general public).

These three steps constitute a *knowledge-based environmentalism*. This has been the working model of environmentalism since the start of the conservation movement in the early part of the twentieth century. It is rooted in the assumptions and methodologies of the knowledge-based worldview as I have previously described it. Knowledge-based environmentalism is a paradigm that influences the work of many major environmental groups, such as the Sierra Club, the Nature Conservancy, the National Wildlife Federation, and the Wilderness Society.

Climate change poses a unique challenge for knowledge-based environmentalism. Most other environmental problems show clear and unquestionable signs of their existence; smog, polluted streams and rivers, urban sprawl, species decline, and disappearing open space are all scientifically verifiable and are, to some extent, intuitively understandable to the layperson as well. Climate change, on the other hand, is not visible and is not clearly noticeable for the nonscientist. In other words, climate change does not exhibit signs that can be plainly and simply attributed to its existence. While there is consensus among most scientists that climate change is a real phenomenon, it is only as real to the general populace as the reports from climate experts are perceived to be certain.

Environmentalism has traditionally relied on both scientific data and obvious environmental damage to bring about changes in laws, technologies, and even the beliefs and actions of individuals. For example, when Rachel Carson wrote about the effects of DDT in *Silent Spring,* changes occurred for two reasons: (1) the effects were scientifically verifiable and widely supported by data; and (2) the solution required technology and policy changes only; that is, it did not bring about a widespread change in the daily activities of average Americans. Carson's book did not demand that we make any sacrifices; instead, it called for eliminating the use of a specific kind of agricultural pesticide. While it might have made the produce in grocery stores safer, *Silent Spring* actually did little to disturb the status quo of American life.

Climate change is far different, not only from DDT, but also from air and water pollution and even ozone hole depletion. While these can all be mitigated by eliminating their corresponding pollutants, climate change differs greatly in scale and in the entrenchment and multifarious nature of its causative factors. Eliminating or substantially reducing worldwide greenhouse gas emissions is a gargantuan task, requiring far more than mandating smokestack "scrubbers" and catalytic converters (air pollution), proper runoff management (water pollution), or the reduction or elimination of chlorofluorocarbons (ozone hole depletion). It should be noted that these "simpler" problems have certainly not been solved by knowledge-based environmentalism: a majority of American produce is grown using pesticides that have not been proved safe, air pollution and water pollution are still significant issues for thousands of Americans, and the ozone hole continues to grow, despite the Montreal Protocol, which mandated the phasing out of chlorofluorocarbons by 1996.

Knowledge-based environmentalism relies on a concrete connection between knowledge and positive change that simply does not seem to exist, perpetuating a flawed assumption that increased knowledge always and necessarily results in increased impetus for "right" action. This faith in the power of knowledge can be easily observed in almost any call for action in response to new findings concerning climate change. Take again the example of the 2007 report of the Intergovernmental Panel on Climate Change, which declares, with "very high confidence," that climate change is real and anthropogenic. On the same day the report was released, the Sierra Club issued a call to arms, emphatically stating that the report gives clear reason to act now, lest we suffer the "dire consequences" of climate change.[12] The underlying assumption here is that nearly definitive and unanimous scientific evidence should be an obvious reason to take action; it is logical that we should act from this "good" science.

There are problems with this line of thinking, however. While climate scientists are in near unanimity concerning the existence and potential severity of climate change, lawmakers in the United States have done little to act on this information, failing even to ratify the Kyoto Protocol, a debatable, yet politically important, step toward stemming climate change. As G. A. Bradshaw and Jeffrey G. Borchers point out in their essay "Uncertainty as Information," the translation of science into policy is unpredictable and tenuous, not deterministic, and far from guaranteed. In addition, they claim that "the science-policy gap may be described in many cases as social inertia born not of a paucity of information but of a deep-seated resistance to change derived from numerous social, religious, and cultural sources."[13] This points to a fundamental problem with assuming that knowledge leads to changing minds and actions. Forces that seem to trump scientific knowledge are in place and persevering.

We must, therefore, consider that knowledge-based environmentalism is inherently limited in its approach to the unwieldy and far-reaching problem of climate change. Considering the delayed effects of greenhouse gas emissions and, most important, that these emissions are tightly and inextricably linked to all aspects of our modern, consumptive lives, it is very possible that knowledge-based environmentalism is not well suited to the difficulties of the climate change phenomenon.

If we continue to approach climate change solely from the knowledge-based perspective, we will be tempted to create false boundaries around it—false because our reductive models of inquiry cannot possibly encompass the boundless complexities of climate change. These boundaries inform our beliefs about the problem and dictate how we respond to it; if our boundaries are wrong, so too will be any proposed "solution" coming from our inquiries within them. Instead, we must expand boundaries to bring forth a better understanding of the true causes of, and the full range of possible solutions to, climate change.

Narrowly classifying climate change as "environmental" has radically limited the search for solutions and allowed us to disregard the fact that it is an issue that cuts across all areas of human experience. Our tendency to do so is atavistic, engendered by the optimism of the 1970s, when knowledge-based environmentalism made notable progress in its push for cleaner technology and progressive legislation. But we should not expect environmental groups to solve what is, ultimately, a complex social, ethical, economic, and political issue. Furthermore, if knowledge-based environmentalism has proved to be less than completely effective when applied to the difficult (but not complex) problems of pollution and ozone depletion, why should we look to it to solve climate change?

It seems clear that we need a new perspective on climate change. That is, we must create a framework for treating it, not just as a scientific or a policy problem, but as a result of the myriad decisions each of us makes every day. This new perspective should be primarily concerned with the task of guiding these decisions and must not rely too heavily on the tenuous link between "right" knowledge and right action.

TOWARD A NEW PERSPECTIVE ON CLIMATE CHANGE

Our hubristic view of scientific knowledge diminishes our respect for the largely unknowable aspects of nature. By assuming that our knowledge about the natural world is now, or will someday be, sufficient, we delusively adopt an authoritative stance over the unknown. Wendell Berry makes the potential for harm in this assumption clear: "To call the unknown 'random' is to plant the flag by which to colonize and exploit the known. . . . To call the unknown by its right name, 'mystery,' is to suggest that we had better respect the possibility of a larger, unseen pattern that can be damaged or destroyed."[14] In the case of climate change, our arrogance has endangered the mysterious intricacies of the natural systems on which all life depends. The time has come to acknowledge and accept the limits of our scientific knowledge.

We would do well to abandon our unfounded faith in the ability of science to render a correct and true picture of the complexity of nature. Furthermore, it is no longer, and never was, safe to believe that science can repair the damages that have resulted from the applications of our always-incomplete scientific knowledge. I believe that climate change should be the impetus for a new perspective on knowledge, one that challenges our enduring scientific optimism, allowing us to see that there is still, and always will be, a need to respect mystery and the unknowable.

The ignorance-based worldview is best described in Wes Jackson's own words: "We are billions of times more ignorant than knowledgeable and always will be."[15] Jackson's claim should raise doubt about our view of knowledge, especially when we are faced with problems as large, complex, and potentially unknowable as climate change. As discussed earlier, the knowledge-based worldview attacks problems from the perspective of "knowledge equals power." The ignorance-based worldview questions the need for power in the first place, balancing our desire for knowledge with a respect and reverence for mystery, never forgetting the axiomatic imbalance between ignorance and knowledge. As a new perspective on climate change, the ignorance-based worldview holds great promise.

The ignorance-based worldview compels us to shift our efforts from proving climate change to admitting our inability to do so; it posits that scientific proof of the existence and the potential harmful effects of climate change should not be a prerequisite for action. This new perspective alone could radically change the strategies of both believers and nonbelievers alike by nullifying both sides of the argument, trumping both those who rely on scientific certainty to bolster public support for action against climate change and the naysayers who claim we do not yet have enough evidence. Instead, we proceed from a position of ignorance. We admit that we do not have enough information about climate change to accurately predict its consequences but that we *do* know enough to give us reason to change our beliefs and ways of living. In fact, we even stop calling the problem "climate change" and, instead, focus on how "environmental" problems defy the narrow boundaries of scientific inquiry. By default, this perspective expands the search for solutions far beyond the limits of a research paradigm, into the always-messy realm of reality.

A broader definition of climate change should result in a skeptical view of the viability of technological fixes for climate change. Currently, some energy experts are touting "green" transportation, such as hybrid and, someday, hydrogen-based automobiles, as the basis for a new world economy and, therefore, a panacea for climate change. While alternative energy is certainly an important *part* of an effective response to climate change, no proposed "solution" that relies solely on technology is complete or necessarily safe. An ignorance-based perspective prompts us to be concerned about the long-term effects of hydrogen-based cars or, for that matter, any new energy source. While hydrogen advocates claim that water vapor emissions are perfectly safe, an ignorance-based approach would demand that we go much further in our research about the effects of hydrogen—from production to delivery to consumption.

It should be clear that one of the primary implications of an ignorance-based worldview is that we should always maintain a position of precaution. As Wes Jackson explains, those with the power to use knowledge in harmful ways must become "assiduous students of exits" by learning to plan for the undesirable effects of our applications of knowledge.[16] Of course, the most common of these applications is our technology; therefore, we must come to realize that our enthusiastic and shortsighted use of it is often accompanied by unanticipated and damaging consequences. The ignorance-based worldview implies that alternative plans—the "exits"—should be prepared *before* we implement new technology, not after the damage is already done.

Imagine how Jackson's idea might have drastically lessened, or perhaps even prevented, climate change! By taking the time to consider that it might not be wise to base our way of life on the necessity of burning fossil fuels, we would have had alternatives—of both energy and lifestyles—in mind and at the ready decades ago. We might have decided that a diversity, rather than a monoculture, of energy and ways of using it was advantageous and more in line with the natural order of things.

Furthermore, under an ignorance-based worldview, action—technological, political, or otherwise—is supplemented with a humbling sense of the tenuousness of our knowledge. While the ignorance-based worldview does not call for us to "absolutize any principle of caution" and take a stance of "radical inaction," it does ask that we always realize that we are operating in a space "between knowledge and ignorance," a space that requires us to be ever mindful of the unknown consequences of both our actions and lack of action.[17]

In short, and perhaps most important, the ignorance-based worldview realizes a fundamental problem when we rely so heavily on science to address climate change: correct actions do not necessarily follow correct discoveries. Given this limitation of knowledge and the potential severity of the climate change problem, we must acknowledge our uncertainties and proceed from a position of ignorance. When we do so, our perspective on climate change is poised for change. This transformation can, and should, come from a respect for what we do not know, perhaps more so than from what we *do* know.

Ignorance-Based Environmentalism

If I have done my job well, I have put forth a convincing argument that an ignorance-based worldview is worth further consideration. The next question that must be posed is: How can the ignorance-based worldview be translated into an environmentalism that is well suited to help us cope with the challenges and uncertainties of global climate change?

First, it would help to acknowledge that environmentalism is much more than the organized groups of specialists who work on a common set of problems loosely or directly related to the natural world. It is also the set of beliefs that inform our perceptions of the "environment": the complex entity that envelops our entire human experience. Therefore, ignorance-based environmentalism is, first and foremost, an expansion in the way problems of the "environment" are viewed, understood, and responded to. It constantly reminds us that these are not environmental problems—these are our problems.

An ignorance-based environmentalism focuses on shifting the faulty perspectives and worldviews at the roots of climate change, but it does not claim to be able to do this on a large-scale level. It is unlikely that a single, top-down catalyst can bring about a worldview shift among the American populace, whether it is as forceful and obvious as a cataclysmic hurricane or as distant and innocuous as the slogans of the mainstream environmental organizations. The kind of deep change of perspective that is required in America will have to take root in local places—the places where we live, work, and play—first. Such change should incorporate a deep sense of skepticism about the power of knowledge and a high regard for the complexity and uncertainty inherent in the problems we face today. An ignorance-based environmentalism excels at making connections between what is done locally and the effects that these local actions have on the global environment. It starts with E. F. Schumacher's dictum ("Think globally, act locally") and ends with a deep awareness of how we must base our actions on a respect for ignorance, instead of optimistically relying on our always-limited and insufficient knowledge.

Ignorance-based environmentalism borrows liberally from civic environmentalism.[18] Civic environmentalism is an alternative to the top-down, technopolitical tactics of mainstream environmentalism, engaging and involving citizens in environmental issues by suggesting that we should think locally and act locally. It keeps the scale small, where local people have the most to lose and the most to gain through their actions. Civic environmentalism holds great promise for addressing noncomplex and localized environmental problems, such as brownfield mitigation and open space conservation. It requires that communities, and the individuals living in them, become aware of what is being done in their local environments. It discourages apathy and demands the creation of, and participation in, a civic life. Moreover, civic environmentalism makes clear the importance of understanding the local nature of environmental problems. It even calls into question the very nature of an "environmental" problem, noting that these problems are inextricably linked to societal and economic issues.

However, if civic environmentalism is to be an effective response to complex environmental issues like climate change, it will need to go further in its inquiry into the nature of environmental problems. It must demonstrate a need to act *regardless of the existence of local degradation.* If it cannot do this, it cannot deal effectively with complex, imperceptible, yet potentially devastating environmental issues. Of course, this is an incredibly challenging task. Human nature is such that change rarely occurs unless it is in response to formidable challenges to the status quo. Obvious and troubling environmental degradation is just such a challenge. Climate change does not widely exhibit such characteristics, however, especially in the continental United States, where reform is most needed. This is precisely why we need an ignorance-based environmentalism, one that does not require observable environmental degradation as an impetus for action. Ignorance-based environmentalism considers the current, nearly unified state of climate science to be more than enough reason to take the precautionary steps of reducing greenhouse gas emissions at home. It does not need local or regional manifestations of climate change before taking the steps to do so.

Climate change is unusually challenging because it is a problem that originates in local places but has global consequences; it is caused by a complex accumulation of innumerable individual actions. Each of us is responsible for causing climate change: while we find it easy to blame the corporations that mass-produce gas-guzzling sport-utility vehicles, it is we who buy them and drive them far too much. While we criticize our national leaders, who are, apparently, not interested in addressing our nation's oil addiction, it is we who fail to insulate our houses better, walk, bike, or carpool more, or do without our yearly vacation to distant places. Each of us, by our modern existence, is responsible for the perpetuation of climate change; each of us is part and parcel of the problem.

However, if we could find ourselves civically engaged and aware in our local places—and mindful of our ignorance—simple, yet powerful, changes might begin to take shape. We might each begin to make one or two decisions differently during each day, decisions based on an awareness of our ultimate inability to predict and control the outcomes of our actions. The exact decisions cannot be determined or predicted; it is not possible to talk with any certainty about how an ignorance-based environmentalism might influence individuals. It is possible, however, to state that, as a framework for decisionmaking, knowing and acknowledging our inescapable ignorance will tend to help us make decisions that are less destructive to both our local and our global environments.

And this, after all, should be the ultimate goal of any environmentalism, whether mainstream, knowledge based, ignorance based, or civic.

A Billionfold Gap: Acting from Ignorance

As I have argued, the ignorance-based worldview is well suited to helping us foster a new environmentalism appropriate for the unusual and challenging nature of climate change. In short, when complete knowledge is not possible, respect for ignorance becomes all the more valuable. In the case of climate change, complete and timely knowledge simply is not possible; the proof of this becomes clear when we consider that greenhouse gas emissions from decades ago are now believed to be contributing to forced climate change. Climate change cannot be undone with technology, and further damage cannot be prevented until and unless we temper our technological and scientific optimism with a respect for ignorance.

It should be clear that there is no simple, universal remedy for climate change; the ignorance-based worldview never assumes that large solutions exist for large problems. It keeps the focus on the small scale yet never forgets the role each of us plays in the greater aspects of the living world. It assumes that our actions have consequences that extend much further than we can ever imagine. By always making this assumption, the ignorance-based worldview is precautionary, slow moving, and self-in-world aware.

Wes Jackson's claim—that there is now, and always will be, a billionfold gap between our ignorance and our knowledge—reminds us that we are living in an immensely complex natural world. This, combined with the difficulties that we have created with our technology, makes one wonder whether there is any reason for hope as we face the real, deep-rooted challenges of climate change. There is a certain paralyzing effect inherent in the ignorance-based worldview if one accepts Jackson's assertion as truth. However, if we take our cue from Jackson himself and the work he is doing at the Land Institute on developing sustainable agriculture, we see that we can accomplish plenty, even in the face of our billionfold ignorance. Finally, when we temper our optimistic view of the power of knowledge, we are left with a profound and necessary sense of humility and a great and well-grounded hope for the future.

NOTES

1. See Bill Vitek, "Joyful Ignorance and the Civic Mind" (in this volume).

2. Francis Bacon, *Novum Organum* (1620), in *The Works,* ed. Basil Montague (Philadelphia: Parry & MacMillan, 1854), 3:343–71, 351, 357, 344, available online at http://history.hanover.edu/texts/Bacon/novorg.html.

3. Ibid., 343, 345–46, 363.

4. Ibid., 367.

5. Ibid., 371.

6. Ibid., 345.

7. Antonio Perez-Ramos, "Bacon's Forms and the Maker's Knowledge Tradition," in *The Cambridge Companion to Bacon,* ed. Markku Peltonen (Cambridge: Cambridge University Press, 1996), 329.

8. Brian Vickers, "Bacon and Rhetoric," in ibid., 496.

9. John Dewey, "The Supreme Intellectual Obligation," *Science,* 16 March 1934, 241.

10. Bacon, *Novum Organum,* 369.

11. United Nations, Intergovernmental Panel on Climate Change, "Climate Change 2007: The Physical Science Basis: Summary for Policymakers" (February 2007), http://www.ipcc.ch/SPM2feb07.pdf.

12. "Statement on IPCC Report from Carl Pope, Sierra Club Executive Director," press release, Sierra Club, 2 February 2007, http://www.sierraclub.org/pressroom/releases/pr2007-02-02.asp.

13. G. A. Bradshaw and J. G. Borchers, "Uncertainty as Information: Narrowing the Science-Policy Gap," *Ecology and Society* 4, no. 1, available online at http://www.ecologyandsociety.org/vol4/iss1/art7.

14. Berry quoted in Wes Jackson's essay "Toward an Ignorance-Based Worldview" in this volume.

15. Berry quoted in Wes Jackson, "Toward an Ignorance-Based World View," *Land Institute,* 3 October 2004, available online at http://www.landinstitute.org/vnews/display.v/ART/2004/10/03/42c0db19e37f4?in_archive=1.

16. Ibid.

17. Steve Talbott, "Toward an Ecological Conversation," *Land Institute* (3 October 2004), 3, available online at http://www.landinstitute.org/pages/ignorance/talbott.pdf.

18. For an introduction to civic environmentalism, see William A. Shutkin, *The Land That Could Be: Environmentalism and Democracy in the Twenty-first Century* (Cambridge, MA: MIT Press).

CAN WE SEE WITH FRESH EYES?

Beyond a Culture of Abstraction

Craig Holdrege

The problem with biases is that we often don't know we have them and aren't aware of how strongly they inform the way we view and act in the world. I want to address one fundamental bias that infects modern Western culture: the strong propensity to take abstract conceptual frameworks more seriously than full-blooded experience. We so easily speak of the world in terms of genes, molecules, atoms, quarks, neural networks, black holes, survival strategies, or other abstract concepts. These are felt to be more "real" than the phenomena of nature we experience—the radiant, blue shimmering Sirius in the winter sky or the deep blue chicory flower that opens at sunrise and fades away before noon.

I suggest that, the more we place abstractions between ourselves and what we encounter in the world, the less firmly rooted we become in that world. The maize that feeds our cattle, pigs, and chickens— grown on immense fields of the Midwest, dowsed with fluid fertilizers that contaminate wells and contribute to oxygen deprivation and death in the lower water layers of the Gulf of Mexico—is much more than a nutrient-generating genetic program modified by human artifice. Viewing maize in such restricted abstract terms, isolated from its larger reality, is what leads us to overlook—at least for a time—the "unfortunate side effects" of our approach. Is it any wonder that a culture caught in a web of abstraction becomes a culture disconnected from nature and destructive in its actions?

In this essay, I want to show some ways to move beyond a culture of abstraction. Since the first step in overcoming a firm habit of mind is to acknowledge its existence, I will call attention to the problem of abstraction itself. Then I will describe how we can open up our perceptual field by trying to put the conceptual element in the background. This entails acknowledging our ignorance and maintaining an ongoing sense of ignorance—and, thereby, intellectual modesty—in all our undertakings. Finally, since we cannot do without concepts, we also have to work on transforming them. This demands changing not only the content of our concepts but also their form or style. I will describe how we can develop what I call living concepts through which we can become more connected to the rich fabric of the phenomenal world.

CAPTURED BY ABSTRACTIONS

The capacity to abstract is what allows us to pull back from our perceptions and look at the world as if from a distance. We can form clear and distinct conceptions about things, make judgments, and then act. In this respect, the ability to abstract is a central feature of being human. But, like all gifts and strengths, our capacity to form abstract concepts is a double-edged sword when it becomes too dominant and habitual. If we do not consciously attend to how we form abstractions and then remain aware of their relation to experience, they tend to take on a life of their own. As a result, we run the danger of attending more to the abstractions themselves than to the world they are meant to illuminate. In this essay, I focus on this shadow side of abstraction.

Here is an extreme description of the world in terms of abstractions by the contemporary philosopher Paul Churchland: "The red surface of an apple does not look like a matrix of molecules reflecting photons at certain critical wave lengths, but that is what it is. The sound of a flute does not sound like a compression wave train in the atmosphere, but that is what it is. The warmth of the summer air does not feel like the mean kinetic energy of millions of tiny molecules, but that is what it is" (Churchland 1988, 15). For Churchland, "reality"—the "is-ness" of things—consists of the high-level abstractions of science. The apples we see and taste, the melody we hear, and the warmth we sense are all only appearances, mere subjective semblances of true physical reality.

And what about our own inwardness? The neuroscientist Antonio Damasio, writing in *Nature,* has an answer: "An emotion, be it happiness or sadness, embarrassment or pride, is a patterned collection of chemical neural responses that is produced by the brain when it detects the presence of an emotionally competent stimulus" (Damasio 2001, 781). So, on this view, the world we experience—all the colors and sounds, smells and tastes—are phantoms of moving molecules, and the joy of eating juicy grapes is "in reality" a chemical response of the brain. This way of viewing things is widely pervasive in science, science education, and science journalism. In one way or another, it comes to inform the way most people today learn to think about the world.

When we raise abstractions onto a pedestal as "the primary realities," we have forgotten how such concepts arise. Concepts such as *molecule, atom,* or *chemical neural responses* develop as the thinking human mind questions the phenomenal world and interacts with it through the experimental method. These concepts are woven out of a rich fabric of theory and experience. When we focus our attention only on the end result, isolated from the rest of the process, we end up with thinglike concepts of atoms and molecules. The problem is that scientific training does not teach us to pay attention to how concepts are formed. Rather, since we usually learn them as abstractions already separated from their genesis—from their actual scientific and human context—we view them as if they were objectlike facts of the world, more real than everything else because they can be so clearly conceived of.

This essentially unconscious process of reification is what the philosopher Alfred North Whitehead called the fallacy of misplaced concreteness (Whitehead 1967, 51ff.; see also Donnelley, in this volume). We treat our abstractions as concrete things of the world. I simply call it object thinking—thinking of the world in terms of objects (Holdrege 1996). The way most people—including scientists who could know better—talk about genes, molecules, hormones, or brain function reveals such object thinking.

So what's the problem with such a way of viewing the world? First of all, it erroneously suggests that, when scientists talk about the world-as-abstraction, they are talking about the world as a whole. What we actually experience—which is not molecules, genes, and firing neurons—becomes a subjective phantasm: the blue of chicory is "only" a particular light wave, water is "only" H_2O, your feelings are "only" your hormones busily at work. Why, in the long run, should we take interest in a world that is "only"? What moral commitment can I have to genes, molecules, and hormones? So one problem with the abstract worldview is that it disconnects us from the very world it sets out to explain.

As the physicist and educator Martin Wagenschein emphasizes, we all too easily ignore the fact that to take a reduced view of the world is a choice (Wagenschein 1975). Physicists have made the choice to view everything in terms of quantities and to mathematize the phenomena. Geneticists have chosen to view heredity in terms of particulate causal entities ("genes"). What these sciences end up with is not a description of the world but a description of one aspect of the world in highly abstract and reduced terms.

As a consequence, conventional modern science and the technologies derived from it address isolated aspects of a much richer fabric of reality. Since this limited perspective of science is often overlooked, we fall into believing that science is addressing *the* problems of *the* world. Nothing is more dangerous than the illusion of thinking you have a solution to a problem (a gene to cure a disease, a pesticide to kill a pest) when you have framed both the problem and the solution in overly narrow terms. Because, however, things play themselves out in complex relations, such solutions may even exacerbate the overall problem (the "cure gene" disrupts other physiological processes; the pests become resistant to the pesticide). As Amory Lovins puts it: "If you don't know how things are connected, then often the cause of problems is solutions" (Lovins 2001).

David Bohm points out that, since scientific concepts and theories lead to a fragmented view of the world (organisms consisting of molecules, molecules consisting of atoms, atoms consisting of elemental particles, etc.), we come to act on the world in a fragmented way:

> If we regard our theories as "direct descriptions of reality as it is," then we will inevitably treat these differences and distinctions as divisions, implying separate existence of the various elementary terms appearing in the theory. We will thus be led to the illusion that the world is actually constituted of separate fragments and . . . this will cause us to act in such a way that we do in fact produce the very fragmentation implied in our attitude to the theory. . . . So what is needed is for man to give attention to his habit of fragmentary thought, be aware of it, and thus bring it to an end. Man's approach to reality may then be whole, and so the response be whole. (Bohm 1980, 9)

Whether we speak of *abstraction, fragmentation, isolation,* or *reductionism* is not so important since each of these terms points to a different nuance of the same habit of mind. What is important is to overcome the habit. If we don't, we will continue to produce myriad unintended effects that inform the ecological, social, and economic problems dominating our times.

THE CONUNDRUM OF KNOWLEDGE

Recognizing the power of abstractions to catch us in their web, the philosopher Edmund Husserl—already nearly a hundred years ago—made an impassioned cry for a "return to the things themselves." But this return—or, perhaps better said, forging ahead—to the things themselves is no easy task, as Husserl describes in *Ideas:*

> That we should set aside all previous habits of thought, see through and break down the mental barriers which these habits have set along the horizons of our thinking . . . these are hard demands. Yet nothing less is required. What makes . . . phenomenology . . . so difficult, is that in addition to all other adjustments a new way of looking at things is necessary, one that contrasts at every point with the natural attitude of experience and thought. To move freely along this new way without ever reverting to the old viewpoints, to learn to see what stands before our eyes, to distinguish, to describe, calls . . . for exacting and laborious studies. (Husserl 1969, 39)

So how can we learn to see with new eyes, to reground our knowing in the world of lived experience rather than in enticing but tenuous abstractions? We can begin by realizing the virtues of ignorance, which is the theme of this collection. Henry David Thoreau describes beautifully in his journals the role of ignorance in knowing:

> It is only when we forget all our learning that we begin to know. I do not get nearer by a hair's breadth to any natural object so long as I presume that I have an introduction to it from some learned man. To conceive of it with a total apprehension I must for the thousandth time approach it as something totally strange. If you would make acquaintance with the ferns you must forget your botany. . . . Your greatest success will be simply to perceive that such things are, and you will have no communication

to make to the Royal Society. (Thoreau 1999, 91 [4 October 1859])

I must walk more with free senses—It is as bad to study stars & clouds as flowers & stones—I must let my senses wander as my thoughts—my eyes see without looking. . . . Be not preoccupied with looking. Go not to the object let it come to you. . . . What I need is not to look at all—but a true sauntering of the eye. (Thoreau 1999, 46 [13 September 1852])

To help us learn this "sauntering of the eye," Thoreau, who was no reticent person, might well have taken us on walks and prodded us with his walking stick to just look, just smell, just hear—and rid ourselves of all our confounded knowledge. But he was also not simpleminded; he knew that there was more involved in knowing:

It requires a different intention of the eye in the same locality to see different plants, as, for example, Juncaceae [rushes] or Gramineae [grasses] even; i.e., I find that when I am looking for the former, I do not see the latter in their midst. . . . A man sees only what concerns him. A botanist absorbed in the pursuit of grasses does not distinguish the grandest pasture oaks. He as it were tramples down oaks unwittingly in his walk. (Thoreau 1999, 83 [8 September 1858])

Thoreau realized that we don't see anything unless we have concepts, unless we have an intention that we bring to the world; otherwise, we would just have confusion. I was once walking along and saw something black moving across the path in front of me. I couldn't "get it." I saw something but had no idea what it was. That was disturbing. I tried the concept *snake,* but it didn't take, and then, suddenly, I saw it: a black plastic bag blowing over the path. The perceptual world, for a moment in disarray, had come together again. Only if I bring concepts to experience do I see coherently.

So there is a problem: the openness and freshness (the ignorance) that allow us to perceive things that don't fit into our preformed ideas and, thereby, to see the unexpected, on the one hand, and the necessity to bring the fruits of previous experience to illuminate the phenomena we are perceiving with openness and freshness, on the other. We need openness to take in something new, but only through applying concepts formed from previous experience—which are, in this sense, biases and can often be quite abstract—can we make sense of the world at all.

So there is a real tension between preformed concepts and openness. I would say that we need to live *actively* and *consciously* within this tension. We need the awareness that gaining knowledge is always a matter of our engaging in the world from a particular perspective. In this way, we become more sensitive to the boundaries of our knowledge and more aware of the extent of our ignorance.

But there is the further question of the quality of our concepts, of what we bring to our experience. Can we transform our concepts so that they become less abstract and more vitally related to experience? Can we move from conceptual biases that color phenomena to more malleable concepts that become sensitive tools to illuminate the not-yet-seen. Can we be just as interested in what does not fit into our scheme of things as we are in what does? Can we continually stretch and remold our view of the world? Or to put it another way: Can we bring new life into our way of knowing?

CULTIVATING OPENNESS

Over a number of years, I studied a particular plant, the skunk cabbage. I was intrigued by its strangeness and wanted to get to know it better. So I went out regularly and observed it, getting to know its habitat, its life cycle, and how it adapted to its environment. I'd often go out with a particular question and focus.

But I also made it a rule to occasionally go out with no fixed focus and try to perceive with Thoreau's "sauntering of the eye." Sometimes it didn't work because my attention would wander inward and I'd start thinking about all sorts of other things. Although I was out in the woods, I was in my head and hardly seeing anything. But sometimes it worked, and I could tell that repeated practice makes it possible to cultivate a kind of open, receptive awareness infused with an animated expectation of what might come toward me.

One March afternoon, I went down to the wetland where skunk cabbage grows. In upstate New York, where I live, it is often still wintry in March. On this day, the sun was shining through the leafless shrubs, and it warmed my face. My eyes were wandering over the skunk cabbage flowers I knew so well that were just emerging from the cool muck. Then I saw a few bees. I watched those bees fly into the flowers and fly out again into other blossoms. In a flash, I realized that I hadn't seen any bees yet that year. The first bees of the year were visiting this plant—this strange plant that warms up to over sixty degrees when it comes out of the ground, even though the air temperature is often at or below freezing. Skunk cabbage warms up, and, on a first somewhat warm and sunny afternoon, the bees come.

I'm pretty sure I would have overlooked this wonderful meeting of bee and skunk cabbage had I not been practicing a "sauntering of the eye." I know myself well as a not-so-open observer and as someone who usually has to focus intently to see. But that very focus can prevent me—and, certainly, often does prevent me—from seeing the unexpected. So, by going out purposefully with the broad focus of open expectation, you overcome your limitations and invite the world in.

Another exercise I use to heighten openness is to think back on the day in the evening and ask myself: "What did I experience today that I wasn't expecting?" It can be disheartening to realize how much of what I experienced was actually expected. Biases were supported: the colleague who is usually a jerk was once again a jerk and so on. To cherish those few moments when something new and unexpected appeared, and then to vividly and concretely repicture those experiences to myself, can lead me to cultivate an interest in and sensitivity to the unexpected. So I can reflect back on my troublesome colleague's actions and words that *did not* fit my expectations. I try to create a field of openness. It actually does bear fruit. I can begin to see another person, a landscape, or a social problem—whatever it may be—with fresh eyes.

BEYOND ABSTRACTION TO LIVING CONCEPTS

Most people think giraffes have long necks. I used to teach, as many biology teachers do, about how the giraffe got its long neck through evolution. The giraffe—as long as I considered it solely in terms of the "fact" of its long neck—was a straightforward illustration of how Darwinian evolution via variation and natural selection works. I was disseminating "knowledge," but did this knowledge really illuminate the giraffe?

Later, I studied the giraffe and its neck in more detail. Since I wasn't interested in any particular theory or explanation and just wanted to get to know the giraffe better, I was open to what the wealth of phenomena had to show me. They showed my ignorance and the poverty of the concepts I'd been using. As a result, the concept of the giraffe's "long neck" increasingly became an abstraction to be overcome.

The first step in overcoming this abstraction was to view the neck both within the context of the whole animal and in comparison with other mammals (Holdrege 2005). I discovered that the neck is not the only thing long in the giraffe. The giraffe has very long and straight legs. Its individual foot and leg bones are long, and, since they are arranged more vertically than in other hoofed mammals, the overall leg length is increased significantly. Moreover, the giraffe is the only hoofed mammal that has longer front legs than hind legs. It has a long head, a very long tongue, and long eyelashes too (and, at the other end, in its tail, the longest hairs you'll find in mammals).

Since the giraffe has a markedly short body in relation to its height—a beautiful instance of what morphologists call compensation—both the neck and the legs appear even longer. I realized the giraffe's neck is part of an overall tendency in the animal toward vertical lengthening, especially in the front part of the body. All the limblike parts of the body—the four legs, the neck as a limb for the head, the jaw of the head, and then, of course, the tongue—are long and, through their particular configuration, allow the animal, for example, to reach high into trees to browse.

But all this has consequences. A giraffe is not only concerned with the world from six to sixteen feet up, where it feeds and browses. It sometimes lowers its head to drink and graze. Then it does something quite strange. It must spread its forelegs awkwardly far apart, making it more vulnerable to predators. Only then can its mouth reach earth or water. The giraffe has a manifestly short neck! What other hoofed mammal has a neck so short that it cannot reach the ground without spreading its legs?

So what is the matter of fact about the fact of the giraffe's long (or short) neck? We come back to what I said before that, if a fact is to be more than an isolated abstraction, we need to view it within a context. And, in the case of the giraffe's neck, the context is the organism itself. Morphologically, the long neck is an exemplary feature of its unique body in which all parts speak long and skyward. But, when the giraffe lowers itself to the terrestrial level, its neck becomes short—an expression of the long-legged animal whose neck attaches so far up on the trunk that its head can no longer reach the ground.

When we frame our questions in abstract ways—What is the cause of the giraffe's long neck?—we have already decided that there is *one* cause and that the giraffe's neck *is* long. We have a terribly oversimplified framework in which we study the animal. The trouble is that we usually don't make the effort to view things within their dynamic, changing contexts. There are lots of stories about how characteristics of organisms evolved, but these stories "work" only as long as you treat the beak, the fin, the feather, or the stomach in isolation from the whole animal. So becoming sensitive to how our concepts inform what we see is important. Without this awareness, we end up explaining schemas and not addressing the things themselves.

What we can do is become more playful with our concepts. When I see the giraffe both in terms of its "long neck" and in terms of its "short neck," I overcome a predilection to look at it in just one way and don't get stuck within a too-narrow conceptual framework. And, at the same time, I begin to appreciate more deeply the organism's complexity. To do justice to this complexity I need to take multiple perspectives. I might not end up with a neat, unified explanation of the animal, but at least I have met the richness of the creature rather than having created an abstract phantom.

As the German poet and scientist Goethe remarked: "If we want to achieve a living understanding of nature we must follow her example and become as mobile and flexible as nature herself" (Goethe 2002, 56 [translation mine]). I have come to realize how organisms can teach us about a living, dynamic way of thinking. If I'm willing to pay attention, I can learn from life how to think in a living way. For me, the study of the growth and development of plants has become an especially vivid and rich model for what I could call living thinking.

A growing plant sends roots spreading intimately through the soil, taking in and exchanging with the earth. These are qualities we have when, as sensory beings, we explore and meet the world with fresh eyes. Always growing, always probing, and meeting things anew. We become rooted in the perceptible experiential world.

As a flowering plant grows, it unfolds leaf after leaf (a process that you can see most vividly in annual wildflowers). When the plant grows up toward flowering, the lower leaves die away. So a plant lives by unfolding something very important at that moment, then moves on to make new structures while past forms fall away. What a wonderful guiding image of how we can work with our concepts: instead of falling in love with a particular conception and holding on to it at all costs—object thinking—we could learn to form a concept, use it, and then let it die away as our experience evolves. Our deeply felt sense of our own boundaries and ignorance allows us to keep knowledge alive, open, and growing. A plant shows us what it means to be undogmatic. Or to put it positively: it shows us how to stay dynamic and adaptable.

You can also read the environment by studying a plant's form. A plant develops differently in drier or richer soil, in shady or brighter light. A plant is always in context. If we were to think plantlike, our concepts would stay closely connected to the context they arose from, and, if that context changed, we would drop or metamorphose our ideas to stay within the stream of life.

In practicing this kind of knowing, we can experience ourselves as active, but also receptive, participants in an ongoing, evolving conversation with nature. We participate even as knowers in the world. We are no longer distant onlookers gazing coolly at a world of objectified things. While gaining this reconnection and rootedness in the world is exhilarating, it is not necessarily comfortable. One of the comfortable things about object thinking is that, because we view the world as consisting of things and have taken on the task of getting at the underlying mechanisms, we can manipulate things at will. Science becomes a kind of value-free zone. But, the moment we become aware of the participatory, interactive nature of knowing, everything changes. Entangled in the world at every moment, we know that we bear responsibility for our way of knowing and its externalization in our technologies and actions. A living thinking is a thinking that knows itself as embedded in the world. It is also a thinking that knows that it does not have "the answer."

CONCLUSION

If we are interested in a new kind of culture, then it won't do to simply tweak the old forms. We need a revolution. Just as the scientific revolution has radically changed the way people view and relate to the world over the past four hundred years, so do we now need a new revolution in worldview that increasingly bears fruit over the next four hundred years.

Seeds of this transformation are created every time we catch ourselves considering a problem or a phenomenon through some preformed conceptual lens and then drop that lens and turn back, in openness, to the things themselves. In this act, we acknowledge our ignorance and show ourselves ready to engage in the concrete situation. With heightened awareness, we can begin forming concepts out of interaction with the world rather than imposing them on it. This is living thinking.

Imagine more and more people cultivating this approach—which is modeled after concrete, living phenomena—rather than striving toward ever greater abstraction in thought (the goal of goals being a unified theory of everything). It will be, at first, a quiet revolution, taking root in the minds of individuals and unfolding in small organizations. But what else would we expect from a revolution modeled after plants? They make no great stir as they go about their work of enlivening the world we live in. The shift from abstraction and object thinking to a plantlike dynamic thinking would help us develop the capacities we need to truly root our understanding and our interactions with nature in nature.

REFERENCES

Bohm, David. 1980. *Wholeness and Implicate Order.* London: Routledge & Kegan Paul.

Churchland, Paul. 1988. *Matter and Consciousness.* Cambridge, MA: MIT Press.

Damasio, Antonio. 2001. "Fundamental Feelings." *Nature* 413, no. 6858 (25 October): 781.

Goethe, Johann Wolfgang von. 2002. "Morphologie." In *Goethes Werke,* vol. 1, *Naturwissenschaftliche Schriften I,* ed. Dorothea Kuhn and Rike Wankemüller, 53–63. Munich: C. H. Beck. Originally published in 1817.

Holdrege, Craig. 1996. *Genetics and the Manipulation of Life: The Forgotten Factor of Context.* Great Barrington, MA: Lindisfarne.

———. 2005. *The Giraffe's Long Neck: From Evolutionary Fable to Whole Organism.* Ghent, NY: Nature Institute.

Husserl, Edmund. 1969. *Ideas: General Introduction to Pure Phenomenology.* New York: Collier. Originally published in 1913.

Lovins, Amory. 2001. Interview with Amory Lovins on September 8, 2001 at the Omega Institute for Holistic Studies. Interviewed by Susan Witt, E. F. Schumacher Society.

Thoreau, Henry David. 1999. *Material Faith: Henry David Thoreau on Science.* Edited by Laura Dassow Wells. Boston: Houghton Mifflin.

Wagenschein, Martin. 1975. "Rettet die Phänomene." In *Erinnerungen für Morgen,* 135–53. Weinheim: Beltz.

Whitehead, Alfred North. 1967. *Science and the Modern World.* New York: Free Press. Originally published in 1925.

CONTRIBUTORS

MICHAEL BERNAS received his M.S. in genetics from the University of Arizona. After working in pathology, he moved to the Fred Hutchinson Cancer Research Center in Seattle and later rejoined the faculty at the University of Arizona in the Department of Surgery. He has numerous publications in both basic and clinical lymphology and has received national and international awards for his work. Mike is the executive editor of the journal *Lymphology* and serves on the Executive Committee of the International Society of Lymphology. In parallel with his funded research, he focuses on science education with special emphasis on kindergarten through graduate questioning and authentic student-based inquiry.

WENDELL BERRY is the author of thirty-two books of essays, poetry, and novels. He has worked a farm in Henry County, Kentucky, since 1965. He is a former professor of English at the University of Kentucky and a past fellow of both the Guggenheim Foundation and the Rockefeller Foundation. He has received numerous awards for his work, including an award from the National Institute and Academy of Arts and Letters in 1971, and the T. S. Eliot Award.

PETER G. BROWN holds concurrent appointments at McGill University in the School of Environment, the Department of Geography, and the Department of Natural Resource Sciences. Before coming to McGill, he was professor of public policy at the University of Maryland's Graduate School of Public Affairs, where he founded the Institute for Philosophy and Public Policy as well as the School of Public Affairs. His teaching, research, and service are concerned with ethics, governance, and the protection of the environment. He is the author of two books: *Restoring the Public Trust* and *The Commonwealth of Life*. He is also involved in tree farming and conservation efforts in Maryland, Maine, and Quebec. He is a certified forest producer in Quebec.

PETER CROWN is multimedia collaboratory producer for the Q^3 program "On Medical Ignorance" at the University of Arizona College of Medicine. He specializes in the development and application of interactive media and uses of the Internet for the Curriculum on Medical (and Other) Ignorance. He received his undergraduate degree from Franklin and Marshall College and his doctoral degree from the University of Arizona. He has worked in academe, business, and public broadcasting, including National Institute of Mental Health research in psychopharmacology and viewer research for *Sesame Street,* was research director for the TV Lab at WNET (PBS), and served as president of two video and multimedia production companies in New York City and Tucson.

RAYMOND H. DEAN is professor emeritus, electrical engineering and computer science, University of Kansas. In retirement, he is promoting energy conservation and the use of renewable energy sources. Before teaching at the University of Kansas, he was the CEO of a heating, ventilation, and air-conditioning and energy-management business. Before that, he did solid-state electronics research at RCA Laboratories. He has a Ph.D. from Princeton University and an M.S. from MIT, both in electrical engineering. He has published seventeen refereed papers in scientific journals and has twenty-one U.S. patents in semiconductor devices and HVAC systems and controls. He is a senior member of the Institute of Electrical and Electronics Engineers.

STRACHAN DONNELLEY is founder (in 2003) and president of the Center for Humans and Nature. Prior to founding the center, he was a past president of the Hastings Center (a bioethics institute) and the director of its former Humans and Nature program. Besides numerous published articles in philosophy and applied ethics, Donnelley has coedited and written for three special supplements to the *Hastings Center Report:* "Animals, Science, and Ethics" (1990), "The Brave New World of Animal Biotechnology" (1994), and "Nature, Polis, Ethics: Chicago Regional Planning" (1999). He also edited a special edition on the philosopher and ethicist Hans Jonas, also in the *Hastings Center Report* (1995). Recently, he has written several articles on philosophy, evolutionary biology, and ethical responsibility.

PAUL G. HELTNE began his career in primate biology and conservation. He accepted in 1982 the position of director, and later president, of the Chicago Academy of Sciences. Becoming president emeritus of the academy in 1999, he has continued to work on civic and scientific projects in the Chicago region and elsewhere in the United States as a director of the Center for Humans and Nature.

CRAIG HOLDREGE, a biologist and educator, is director of the Nature Institute in rural upstate New York. The Nature Institute is dedicated to research and educational activities applying holistic and phenomenological methods (see www.natureinstitute.org). Holdrege is keenly interested in the interconnected nature of things and carries out holistic studies of plants and animals. He also critically examines new developments in genetics and biotechnology from a contextual perspective. He is the author of *The Giraffe's Long Neck: From Evolutionary Fable to Whole Organism* and *Genetics and the Manipulation of Life: The Forgotten Factor of Context.*

WES JACKSON is president of the Land Institute. He established and served as chair of one of the country's first environmental studies programs at California State University, Sacramento, and then returned to his native Kansas to found the Land Institute in 1976. He is the author of several books, including *New Roots for Agriculture* and *Becoming Native to This Place,* and is widely recognized as a leader in the international movement for a more sustainable agriculture. He was a 1990 Pew Conservation Scholar, in 1992 became a MacArthur Fellow, and in 2000 received the Right Livelihood Award (called the "alternative Nobel Prize").

JON JENSEN is assistant professor of philosophy and environmental studies at Luther College in Decorah, Iowa. He received a Ph.D. in philosophy from the University of Colorado and taught in Vermont before returning to the Midwest, where he has deep roots. He is the coauthor of *Questions That Matter: An Invitation to Philosophy* as well as articles in environmental philosophy. His current research focuses on connections between sustainable agriculture, ecological restoration, and local food systems.

RICHARD D. LAMM is codirector of the Institute for Public Policy Studies at the University of Denver and a former three-term governor of Colorado (1975–1987). He is both a lawyer (University of California, Berkeley, 1961) and a certified public accountant. He joined the faculty of the University of Denver in 1969 and has, except for his years as governor, been associated with the university ever since. He has received several awards, appeared on numerous national news programs, and written many editorials as well as writing or coauthoring six books.

CHARLES MARSH is the William Allen White Foundation Professor of Journalism and Mass Communications at the University of Kansas. He has won awards for excellence in teaching from students at the University of Kansas and the Consortium International University in Italy. He recently presented papers on ethics and critical thinking at the annual conferences of the Classical Association of England and Wales and the Association for Education in Journalism and Mass Communication.

JOE MAROCCO holds a bachelor's degree from Union College in Schenectady, New York, and a master of arts degree from Empire State College in Saratoga

Springs, New York. His graduate work, which was completed in the summer of 2006, focused on the philosophical aspects of the global climate change problem. His master's thesis, "Ignorance-Based Environmentalism: Climate Change and the Limits of Knowledge," was inspired by his work with Dr. Bill Vitek of Clarkson University and was the basis for his essay in this volume. He lives, plays, and works in the Adirondack Mountains with his wife and daughter.

CONN NUGENT is executive director of the J. M. Kaplan Fund and oversees grant programs in the environment, historic preservation, and immigration. The Kaplan Fund focuses on New York City and key cross-border regions of North America. Previously, he served as executive director of the Citizens Union of the City of New York, the oldest good-government organization in the United States. From 1989 to 1995, he was the environment program director at the Nathan Cummings Foundation, where he established programs on transportation, agriculture, and tax policy. He is a graduate of Harvard College and Harvard Law School. He has published articles in a variety of leading U.S. publications on topics ranging from architecture to baseball.

ROBERT PERRY is currently an environmental consultant to foundations and the interim director of the Albemarle Ecological Field Site for the University of North Carolina, Chapel Hill. He worked for the Geraldine R. Dodge Foundation for eleven years, first as its education program officer, then as its director of environment and welfare of animals programs. His personal passion centers on the conservation of biological diversity, particularly with regard to the myriad small, often unseen and unsung organisms (beetles, minnows, corals, and frogs) that sustain the larger animals that people know better and, therefore, tend to appreciate more. He taught the sciences in New York City for seventeen years while obtaining his master's degree at the City University of New York. He taught educators for several years through City College's Environmental Studies Program and began doctoral work at Columbia University. He currently lives with his family on Roanoke Island off the coast of North Carolina, where the natural world exists right outside his door.

ANNA L. PETERSON received an A.B. from the University of California, Berkeley, and a Ph.D. from the University of Chicago Divinity School. She is a professor in the Department of Religion at the University of Florida and an affiliate professor in the Center for Latin American Studies and the School of Natural Resources and the Environment. Her research focuses on religion in contemporary Latin American society, especially Central America, and on social and environmental ethics. She has written numerous articles in journals as well as several books, the most recent of which is *Seeds of the Kingdom:*

Utopian Communities in the Americas. Her current research focuses on the gap between environmental values and actual practices.

ROBERT ROOT-BERNSTEIN earned bachelor's and doctoral degrees at Princeton University and was among the first group of MacArthur Fellows. He splits his time between doing science (autoimmune diseases, drug development, and the evolution of physiological control systems) and studying how best to do science (the history and philosophy of science and the nature of the creative process). He is professor of physiology at Michigan State University, where he has won numerous teaching awards.

STEVE TALBOTT worked for many years in the high-tech field and is now a senior researcher at the Nature Institute in Ghent, New York. He writes the online newsletter *NetFuture* (www.netfuture.org), which was termed "a largely undiscovered national treasure" in a *New York Times* feature story on his work. Steve is the author of *The Future Does Not Compute: Transcending the Machines in Our Midst,* named by the academic library journal *Choice* to its list of "Outstanding Academic Books of 1996." His most recent book is *Devices of the Soul: Battling for Our Selves in an Age of Machines.*

HERB THOMPSON received a doctorate in economics from the University of Colorado in 1969. Presently a professor of economics at the American University in Cairo, Egypt, he has also taught at universities in Australia, Papua New Guinea, Alaska, Ohio, and Missouri. Most recently, his published work has concentrated on tropical rain forest deforestation in the Asia/Pacific region and cybersystemic teaching and learning. His more satisfactory personal accomplishments in life include ten years as a professional jazz drummer, learning to play the 'oud (an Egyptian lute), and achieving the purple belt in jiu jitsu.

BILL VITEK is associate professor of philosophy at Clarkson University in Potsdam, New York. His research is focused at the intersection of social practices and their environmental, cultural, and historical contexts. He is the author of the book *Promising* (1993) and numerous articles and essays on the topics of agriculture, environmental ethics, social policy, and civic philosophy. He is coeditor—with Wes Jackson—of *Rooted in the Land: Essays on Community and Place* (1996). He is currently working on a coauthored book with Jackson and Strachan Donnelley titled *Consulting the Genius of the Place* and a book of his own essays titled *Toss the Paddle! Finding Our Way Out of Carbon Creek.* He is also a professional jazz pianist.

CHARLES L. WITTE, M.D., graduated from Columbia College and received his M.D. from New York University. In conjunction with Marlys H. Witte,

M.D., he established the first lymphological laboratory in the United States associated with a comprehensive lymphedema-angiodysplasia diagnostic and treatment center. He was the editor-in-chief of *Lymphology* and the author or coauthor of more than four hundred articles and book chapters on a wide variety of medical subjects. He also joined with Dr. Marlys Witte and the philosopher Ann Kerwin in developing the Curriculum on Medical (and Other) Ignorance. He received international recognition for his landmark studies in lymphology and liver cirrhosis, including election to the National Academy of Medicine of Brazil. He considered surgery to be the ultimate discipline of ignorance—surgeons remove organs because they cannot otherwise prevent or medically treat the diseases affecting them. He passed away on 7 March 2003.

MARLYS HEARST WITTE, M.D., is professor of surgery and director of the Medical Student Research Program at the University of Arizona and is also secretary-general of the forty-two-nation International Society of Lymphology. She specializes in disorders of the lymphatic system (lymphatics, lymph nodes, lymph, and lymphocytes) and conducts research in basic and clinical lymphology. She received her undergraduate degree from Barnard College, her medical education at the New York University School of Medicine, and her postgraduate residency training at the University of North Carolina, North Carolina Memorial Hospital, and New York University, Bellevue Medical Center. Together with colleagues, including her late husband, Charles L. Witte, M.D., a practicing surgeon, educator, and researcher, she has developed the internationally recognized Curriculum on Medical (and Other) Ignorance for students at all levels.

INDEX